AF845306

Novel Techniques for Investigating Recharge to the Memphis Aquifer

Awwa
Research
Foundation

About the Awwa Research Foundation

The Awwa Research Foundation (AwwaRF) is a member-supported, international, nonprofit organization that sponsors research to enable water utilities, public health agencies, and other professionals to provide safe and affordable drinking water to consumers.

The Foundation's mission is to advance the science of water to improve the quality of life. To achieve this mission, the Foundation sponsors studies on all aspects of drinking water, including supply and resources, treatment, monitoring and analysis, distribution, management, and health effects. Funding for research is provided primarily by subscription payments from approximately 1,000 utilities, consulting firms, and manufacturers in North America and abroad. Additional funding comes from collaborative partnerships with other national and international organizations, allowing for resources to be leveraged, expertise to be shared, and broad-based knowledge to be developed and disseminated. Government funding serves as a third source of research dollars.

From its headquarters in Denver, Colorado, the Foundation's staff directs and supports the efforts of more than 800 volunteers who serve on the board of trustees and various committees. These volunteers represent many facets of the water industry, and contribute their expertise to select and monitor research studies that benefit the entire drinking water community.

The results of research are disseminated through a number of channels, including reports, the Web site, conferences, and periodicals.

For subscribers, the Foundation serves as a cooperative program in which water suppliers unite to pool their resources. By applying Foundation research findings, these water suppliers can save substantial costs and stay on the leading edge of drinking water science and technology. Since its inception, AwwaRF has supplied the water community with more than $300 million in applied research.

More information about the Foundation and how to become a subscriber is available on the Web at **www.awwarf.org**.

Novel Techniques for Investigating Recharge to the Memphis Aquifer

Prepared by:
Randy Gentry, **Larry McKay**, and **Norbert Thonnard**
University of Tennessee, Knoxville, TN 37996

Jerry L. Anderson and **Daniel Larsen**
University of Memphis, Memphis, TN 38152

John K. Carmichael
U.S. Geological Survey, Nashville, TN 37211

and

Kip Solomon
University of Utah, Salt Lake City, UT 84112

Jointly sponsored by:
Awwa Research Foundation
6666 West Quincy Avenue, Denver, CO 80235-3098

and

Memphis Light, Gas and Water
Memphis, TN 38103

Published by:

Awwa Research Foundation

American Water Works Association

IWA Publishing

DISCLAIMER

This study was jointly funded by the Awwa Research Foundation (AwwaRF) and Memphis Light, Gas and Water (MLG&W). AwwaRF and MLG&W assume no responsibility for the content of the research study reported in this publication or for the opinions or statements of fact expressed in the report. The mention of trade names for commercial products does not represent or imply the approval or endorsement of AwwaRF or MLG&W. This report is presented solely for informational purposes.

John K. Carmichael of U.S. Geological Survey (USGS) authored Chapter 2 and portions of Chapter 5. USGS assumes no responsibility for the content of the research study reported in these chapters, or for the opinions or statements of fact expressed in them. The mention of trade names for commercial products in Chapters 2 and 5 does not represent or imply the approval or endorsement of USGS.

CONTENTS

TABLES

FIGURES

FOREWORD

The Awwa Research Foundation (AwwaRF) is a nonprofit corporation that is dedicated to the implementation of a research effort to help utilities respond to regulatory requirements and traditional high-priority concerns of the industry. The research agenda is developed through a process of consultation with subscribers and drinking water professionals. Under the umbrella of a Strategic Research Plan, the Research Advisory Council prioritizes the suggested projects based upon current and future needs, applicability, and past work; the recommendations are forwarded to the Board of Trustees for final selection. The foundation also sponsors research projects through an unsolicited proposal process; the Collaborative Research, Research Applications, and Tailored Collaboration programs; and various joint research efforts with organizations such as the U.S. Environmental Protection Agency, the U.S. Bureau of Reclamation, and the Association of California Water Agencies.

This publication is a result of one of these sponsored studies, and it is hoped that its findings will be applied in communities throughout the world. The following report serves not only as a means of communicating the results of the water industry's centralized research program but also as a tool to enlist the further support of the nonmember utilities and individuals.

Projects are managed closely from their inception to the final report by the foundation's staff and large cadre of volunteers who willingly contribute their time and expertise. The foundation serves a planning and management function and awards contracts to other institutions such as water utilities, universities, and engineering firms. The funding for this research effort comes primarily from the Subscription Program, through which water utilities subscribe to the research program and make an annual payment proportionate to the volume of water they deliver and consultants and manufacturers subscribe based on their annual billings. The program offers a cost-effective and fair method for funding research in the public interest.

A broad spectrum of water supply issues is addressed by the foundation's research agenda: resources, treatment and operations, distribution and storage, water quality and analysis, toxicology, economics, and management. The ultimate purpose of the coordinated effort is to assist water suppliers to provide the highest possible quality of water economically and reliably. The true benefits are realized when the results are implemented at the utility level. The foundation's trustees are pleased to offer this publication as a contribution toward that end.

Walter J. Bishop
Chair, Board of Trustees
Awwa Research Foundation

Robert C. Renner, P.E.
Executive Director
Awwa Research Foundation

ACKNOWLEDGMENTS

The investigators would like to express appreciation to Mr. Chris Rayburn of AwwaRF and the Project Advisory Committee Drs. Lee Davisson, June Fabryka-Martin, and John Van Brahana for their valuable comments during the implementation of this research plan. Mr. Charles Pickle, now retired from the Memphis Light, Gas and Water division and Dr. Fred Von Hofe were very helpful and supportive. In addition, the Memphis Light, Gas and Water division provided personnel and equipment to aid the researchers during the drilling phase and the collection of samples during wet weather conditions. The investigators especially appreciate Mr. Jeff Embry's help at the field site. It would not have been possible to complete the project without the help of many individuals at the Memphis Light, Gas and Water division. The investigators would also like to express appreciation to Mr. White of White Construction for the use of personnel and equipment for the transportation of water to the drilling sites. Also, Mr. Ron Smith of the Memphis District of the Army Corps of Engineers was responsive and very helpful in the drilling of test holes prior to the implementation of well nest drilling for the project. Many personnel with the University of Memphis Ground Water Institute were helpful in the collection of field samples including Drs. Stephanie Ivey and Brian Waldron, and many graduate students. The investigators appreciate the University of Memphis' College of Engineering for providing space for the archival of drilling samples from the project.

EXECUTIVE SUMMARY

The ability to deliver a safe supply of drinking water to the public is of paramount concern to all public and private utilities. The behavior and especially the potential for contamination of aquifer systems by localized leakage or recharge are not always well understood. An important aspect of managing groundwater resources is the ability to understand the behavior of aquifer systems so that risks to the resource can be better assessed. This study of recharge mechanisms to the Memphis aquifer in Shelby County, Tennessee, sought to develop better tools to enhance MLG&W's capabilities in managing water resources of the Memphis aquifer and improve source-water protection. The specific objective of the study was to develop a better understanding of the magnitude and impact of recharge through windows in the upper Claiborne confining layer.

For the study, the Shelby Farms site was selected on the basis of a previously identified window in the upper Claiborne confining unit at the site. Previous studies have established that groundwater flows laterally within the shallow aquifer to the window in the upper Claiborne confining unit and then downward as recharge to the Memphis aquifer. The groundwater in the shallow aquifer is locally affected by landfill leachate, which provides a chemical tracer of groundwater flow patterns. The experimental design included installation of two clusters of observation wells, one within the window (the unconfined cluster) and the other along the prevailing groundwater flow direction down gradient from the window in a confined part of the Memphis aquifer (the confined cluster). During the week of March 26, 2001, three exploratory test holes were drilled at separate locations at the Shelby Farms study area to (1) refine the western and northwestern extent of the window by determining the extent of confining clay at these locations, and to (2) select the best location for the confined well cluster. The effort was successful in confirming the presence of clay layers to the north and northwest of the window location.

Subsequent to the test hole drilling, installation of the well nests was conducted during the summer of 2001. Nine monitoring wells were constructed in and around the window site for further characterization purposes. The nested sites included three monitoring wells adjacent to an existing monitoring well at the window (unconfined) site, four additional wells at the confined site located downgradient of the window, and two Memphis aquifer wells adjacent to existing shallow aquifer wells located northwest and southwest of the window site. Rotasonic cores were collected from the deepest borings at the confined and unconfined locations and from selected intervals of the boring at the northwestern well nest. The cores from the southwestern and northwestern well nests were described in detail and sampled at discrete intervals at the site prior to discarding the remainder of the cores. Monitoring wells were completed in each of the newly drilled boreholes. Hydraulic testing of selected wells was conducted during April through June of 2002.

The new wells allowed for a more spatially refined assessment of heads in the shallow and Memphis aquifers near the vicinity of the window. Water levels in all existing and accessible observation and production wells in the vicinity of the Shelby Farms window site were measured on three occasions over the project period (June 12 to 13, 2002; October 24, 2002; May 27 to 28, 2003). These data were used to prepare potentiometric surface maps for the alluvial (shallow) aquifer and the Memphis aquifer for late spring/early summer and fall conditions. These data were also compared to the select measurements of stage and discharge of the Wolf River, which flows near the window site.

In addition to hydraulic evaluation, a detailed sedimentological and mineralogical investigation based upon core samples from the borings was performed. Selected samples were analyzed for their (1) grain-size distribution to better determine the hydraulic conductivity of the sedimentary units, (2) total organic carbon content to better estimate sorptive characteristics, and (3) mineralogy to help determine origin and sorptive characteristics. Based upon the detailed grain-size and mineralogical investigation, a new hydrostratigraphic conceptual model is proposed for the Shelby Farms site. Five hydrostratigraphic units are defined, each with different hydraulic and stratigraphic properties: (1) the upper alluvium, a relatively homogenous unit with low matrix hydraulic conductivity, but numerous root traces and soil-related fractures that may substantially enhance porosity and permeability; (2) the shallow aquifer, a sand and gravel unit with the highest average hydraulic conductivity; (3) confining unit facies A, a dense, dark olive gray silty clay with low permeability and chemically reduced character, (4) confining unit facies B, a less dense, white interbedded sand and clay with intermediate hydraulic properties and chemically oxidized character; and (5) the Memphis aquifer, a relatively high hydraulic conductivity sandy unit. The upper Claiborne confining unit is absent at the unconfined site; thus, the shallow aquifer and Memphis aquifer are in direct contact. The contact between the Memphis aquifer and the confining unit deepens abruptly in all directions away from the unconfined site. The two relationships demonstrate that the window at the confined site is limited in its extent and is largely a result of heterogeneities in stratigraphic units beneath the shallow aquifer.

From the onset of the project, the plan included a scheme to determine the age distribution of the vertical recharge at the window site along with a comparison to down-gradient age distributions. The environmental tracers, tritium, helium (for $^{3}H/^{3}He$ dating) and krypton (^{85}Kr), were to be used along with chemical data for dating and tracing the groundwater. The researchers obtained samples for both sets of environmental tracer analyses, but only the $^{3}H/^{3}He$ samples were analyzed. Although the development of the ^{85}Kr analysis method at the Institute for Rare Isotope Measurements (IRIM) at the University of Tennessee had not yet been completed when this research project was proposed, it was decided to include sampling and analysis for this emerging environmental tracer because of the opportunity to compare results with more established tracers, such as $^{3}H/^{3}He$, in a setting that is particularly applicable to ^{85}Kr. Unfortunately, the analytical method is technically very challenging, and it has become clear that no ^{85}Kr sample results will be available in time for this final project report. The $^{3}H/^{3}He$ results have permitted us to complete the scientific goals of the project, and it was decided not to further delay the project by waiting for the ^{85}Kr analysis method to come on-line.

The geochemical results indicate that landfill leachate has substantially intruded into the Memphis aquifer within and downgradient of the window area. Contaminant concentrations and redox boundaries assist in defining the limits of the leachate plume and indicate that groundwater flows generally from the unconfined site to the confined site, as is suggested from the regional groundwater flow data. Tritium-^{3}He results indicate relatively rapid vertical recharge through the window followed by stratified, but mixed, descending flow toward the confined site. The base of the leachate plume in the Memphis aquifer was not accessed by any of the wells. The pattern of age dates and geochemical data as well as other data suggest leakage may occur at another gap in the confining unit upgradient of the unconfined site.

Numerical modeling of the study area using a steady-state finite difference technique was implemented to determine the likely distribution of recharge to the Memphis aquifer. A genetic algorithm optimization technique was coupled to a finite difference-based forward model to explore the likely distribution of constant recharge to the Memphis aquifer. Subsequently, MODFLOW 2000 was used with a zonation technique to investigate the most likely distribution of non-uniform recharge to the Memphis aquifer.

The study was successful in describing in more detail the flow behavior and unique age distribution characterization at the shallow aquifer – Memphis aquifer interface at the Shelby Farms site. The application of a multitude of techniques to arrive at a hydraulically and geochemically robust conceptual model is considered to be a benefit to the regional community and MLG&W. The combined tools used to study the window site have resulted in a new proposed hydrostratigraphic conceptual model for the site explaining the most probable genesis of the window as well as groundwater flow patterns in and around the window. The study has concluded that the most likely maximum recharge rate through the window is in the range of 0.04 to 0.09 ft/day. The study has demonstrated that confining unit heterogeneities play an important role in the aquifer hydraulics at the site, and that windows play a major role in aquifer recharge, at least at the local scale. Given an improved understanding of aquifer recharge mechanisms and contamination potential developed from this study, local utility decision makers should have better tools to assist in the management of the groundwater resources.

CHAPTER 1
INTRODUCTION AND SCOPE

BACKGROUND

The Memphis aquifer comprises the Memphis Sand of Tertiary age and is the major water resource for Shelby County and western Tennessee (Parks and Carmichael, 1990). The Memphis Aquifer is part of the Mississippi Embayment aquifer system, which extends from southwestern Kentucky to the Gulf of Mexico (Graham and Parks, 1986). The utilities that extract water from the Memphis aquifer in Shelby County include Memphis Light, Gas and Water (MLG&W), the City of Germantown, the Town of Collierville, the City of Bartlett, and the City of Millington. The Memphis aquifer outcrops in eastern and southeastern Shelby County, but the aquifer is confined by the overlying upper Claiborne confining unit (a low permeability clay layer) throughout most of the remainder of the county (Figures 1.1 and 1.2; Table 1.1). Overlying the upper Claiborne confining unit are layers of gravel, sand, and silt which form a shallow unconfined aquifer system (also referred to as the water-table aquifer). The shallow unconfined aquifer system is highly susceptible to contamination from urban, industrial and agricultural activities and is not widely used as a water supply (Parks, 1990; Robinson et al., 1997). The Memphis aquifer is partially protected from contamination by the overlying upper Claiborne confining unit, which is as much as 375 feet thick in western part of the Memphis area.

IMPLICATIONS TO GROUNDWATER RESOURCE MANAGEMENT

For many years, the principal source of recharge to the Memphis aquifer was believed to be infiltration of precipitation in eastern Shelby County and western Fayette County in the outcrop area of the Memphis Sand. However, over the past decade, numerous studies have shown that recharge and contamination also are entering the Memphis aquifer through "windows" or hydraulic openings through the upper Claiborne confining unit, which allow for direct hydraulic communication or leakage between the shallow unconfined aquifers and the deeper Memphis aquifer. Windows can be formed by depositional or erosional processes, as well as by earthquake activity, and may be common in many layered aquifer systems. As groundwater withdrawal from the Memphis aquifer increases and heads in the deeper aquifer are decreased, the potential exists for greater amounts of poorer quality water from the shallower zones to be drawn down through these windows.

Detection of windows and assessing their contribution to aquifer recharge and water quality is difficult and expensive with conventional drilling and testing methods. Locating all of the significant windows in Shelby County based solely on drilling would likely take many hundreds or thousands of boreholes. In some cases, locations of windows have been inferred from observed depressions in the water-table aquifers, or from geochemical evidence of mixing of waters from the shallow aquifers and the Memphis aquifer (Graham and Parks, 1986; Richardson, 1989; Parks and Mirecki, 1992; Parks et al., 1995).

This study has advanced the use of novel techniques for assessing the effects of windows on recharge and water quality using a known window site within a layered aquifer system. A variety of hydraulic, geochemical, and numerical modeling tools have been developed over the

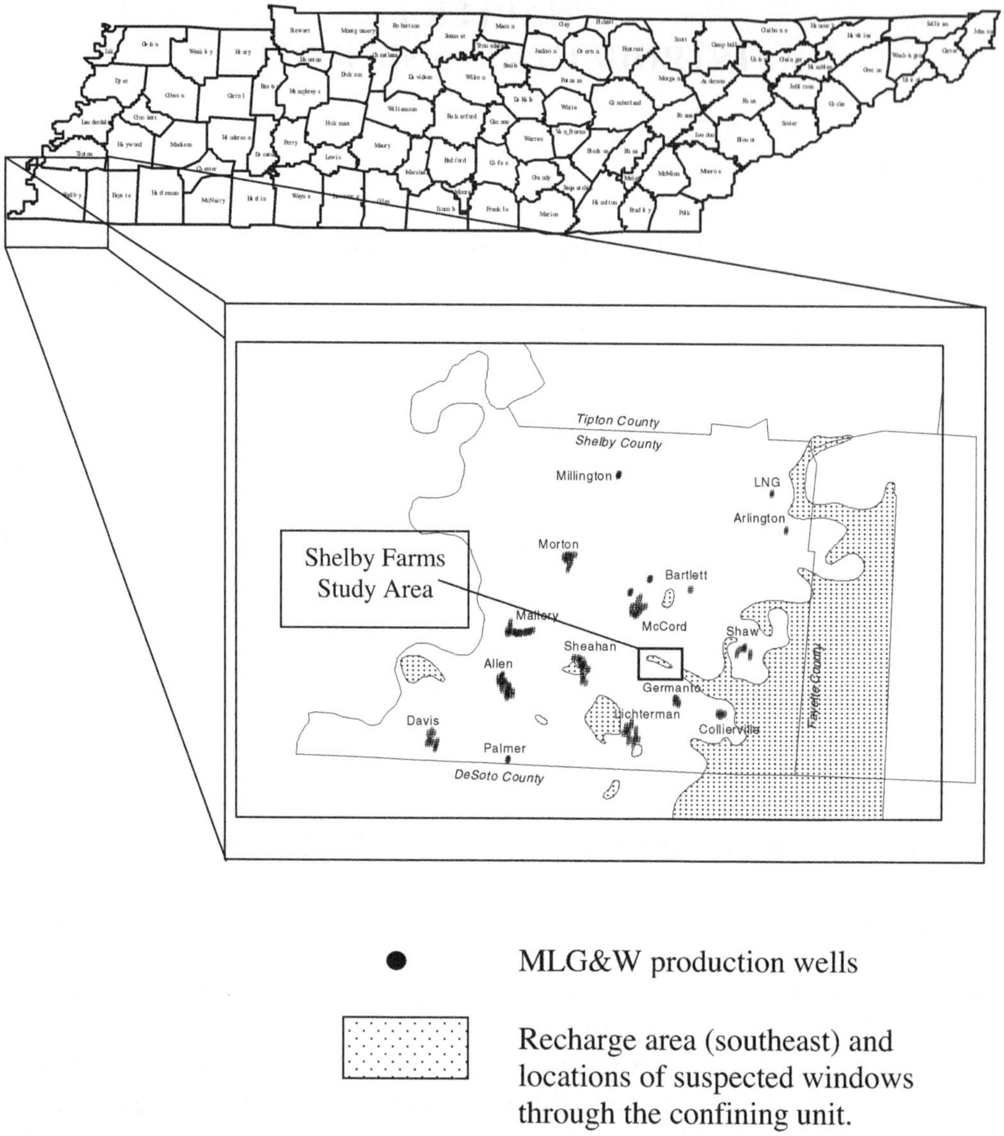

Source: Adapted from Kingsbury and Parks, 1993.

Figure 1.1 Map of Shelby County showing the distribution of the recharge area for the Memphis aquifer and locations of suspected windows through the upper Claiborne confining unit. Also shown are the Memphis Light, Gas and Water and various other municipal well fields and the Shelby Farms study area.

past 20 years to assess aquifer recharge, but most of their applications have either been in very simple flow systems, such as homogeneous unconfined aquifers, or in regional flow systems where parameters are "averaged" over large areas (e.g., Castro et al., 1998ab). Some inverse modeling techniques have recently been used to attempt to identify aquifer recharge fluxes; however, limited data are available for validation of these methods (Gentry et al., 2001). This study was designed to address the need to develop and test integrated procedures involving a combination of hydraulic, geologic, geochemical, and numerical modeling approaches for assessing recharge and contaminant migration in confined aquifer systems.

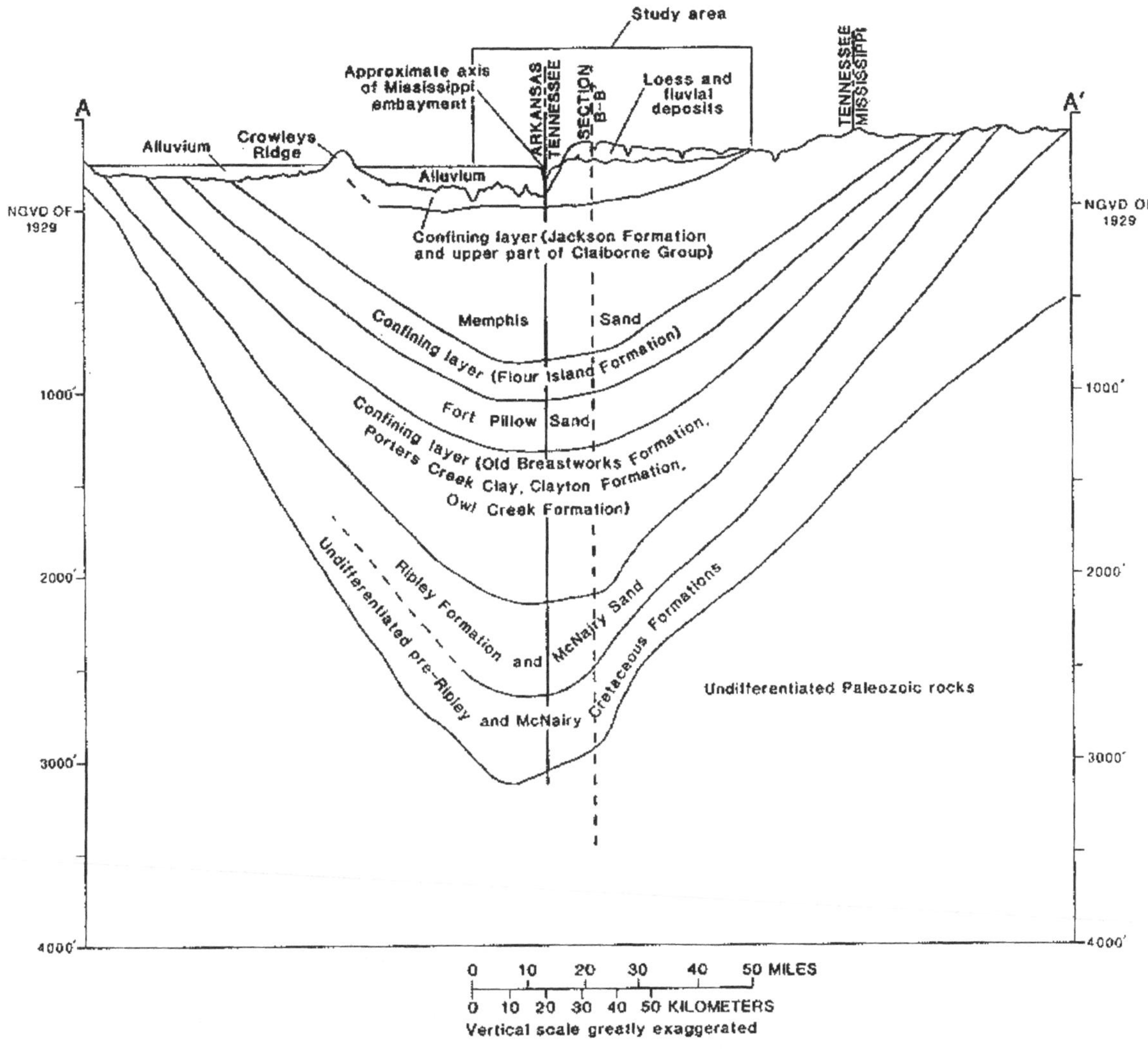

Source: Adapted from Brahana, Parks, and Gaydos, 1987.

Figure 1.2 An east-west geologic cross-section showing the stratigraphy of the northern Mississippi Embayment through Memphis, Tennessee

SCOPE OF STUDY

The results and conclusions of the research project described herein are the result of efforts aimed at better understanding the hydraulics of interaquifer flow between the shallow aquifer system and the Memphis aquifer in the Memphis area. Specifically, the objective of this Tailored Collaboration research project was to develop better analytical tools that will enhance MLG&W's capabilities in managing development of the Memphis aquifer to assure its long-term viability and to protect it from contamination. The objective was achieved by developing a better understanding of the magnitude and impact of leakage through windows in the upper Claiborne confining unit relative to recharge through the confining layer and in the unconfined part of the

Table 1.1
Post-Midway Group geologic units underlying the Shelby County

System	Series	Group	Stratigraphic unit	Thickness (ft)	Lithology and hydrologic significance
Quaternary	Holocene and Pleistocene		Alluvium (alluvial deposits)	0-175	Sand, gravel, silt, and clay. Underlies the Mississippi Alluvial Plain and alluvial plains of streams in the Gulf Coastal Plain. Thickest beneath the Mississippi Alluvial Plain, where it is commonly between 100 and 150 feet thick and makes up the Mississippi River Valley alluvial aquifer. Generally less than 50 feet thick elsewhere. Provides water to domestic, farm, industrial, and irrigation wells in the Mississippi Alluvial Plain.
	Pleistocene		Loess	0-65	Silt, silty clay, and minor sand. Principal unit at the surface in upland areas of the Gulf Coastal Plain. Thickest on the bluffs that border the Mississippi Alluvial Plain; generally thinner eastward. Tends to retard downward movement of water providing recharge to the water-table aquifers (alluvium and fluvial deposits).
Quaternary and Tertiary (?)	Pleistocene and Pliocene (?)		Fluvial deposits (terrace deposits)	0-100	Sand, gravel, minor clay and ferruginous sandstone. Generally underlies the loess in upland areas, but are locally absent. Thickness varies greatly because of erosional surfaces at top and base. Provides water to many domestic and farm wells in rural areas.
Tertiary	Eocene		Jackson Formation	0-50	Sand, silt, clay, and lignite. Because of similarities in lithology, the Jackson and Cockfield Formations cannot be reliable subdivided based on available information. Thicknesses are estimates based on tentative geophysical-log correlations. The Jackson and Cockfield provide water to a few domestic wells in rural areas.
		—?—	—?—	—	
			Cockfield Formation	0-250	
		Claiborne	Cook Mountain Formation	0-110	Clay, silt, and sand. Generally consists of clay and silt, but locally may consist predominantly of fine sand. Probably averages about 60 feet in thickness. Serves as upper confining unit for the Memphis aquifer, along with the Cockfield and Jackson Formations.
			Memphis Sand ("500-foot" sand)	650-900	Sand, silt, clay, and minor lignite. Consists of a thick body of sand with clay lenses at various horizons. Sand is fine to very coarse. Upper part commonly contains lenses of fine sand, silt, and clay; lower part locally contains lenses of clay as thick as 50 feet. The Memphis aquifer—the principal aquifer in the Memphis area—provides water for most domestic, commercial, industrial, and municipal supplies, including the well fields of Memphis Light, Gas and Water Division.
	Paleocene	Wilcox	Flour Island Formation	140-310	Clay, silt, sand, and lignite. Not an aquifer. Consists predominantly of clay and silt, but locally contains lenses of fine sand as thick as 50 feet. Serves as the lower confining unit for the Memphis aquifer and the upper confining unit for the Fort Pillow aquifer.
			Fort Pillow Sand ("1400-foot" sand)	98-300	Sand and minor clay. Sand is fine to medium or medium to coarse. Thickest in the southwestern part of the Memphis area; thinnest in the southeastern part. The Fort Pillow aquifer is the second principal aquifer supplying water for the City of Memphis. Principal aquifer providing water for municipal and industrial supplies west of the Mississippi River.
			Old Breastworks Formation*	180-310	Clay, silt, sand, and lignite. Not an aquifer. Consists predominantly of clay and silt. Serves as the lower confining unit for the Fort Pillow aquifer, along with the underlying Porters Creek Clay and Clayton Formation of the underlying Midway Group.

Source: Adapted from Kingsbury and Parks (1993) and Carmichael et al., (1997).

* Frederiksen and others (1982) tentatively placed the Old Breastworks Formation in the Midway Group. For this report, the Old Breastworks Formation of the Wilcox Group is used as defined by Moore and Brown (1969).

aquifer. The goals of the study, as presented in previous progress reports, are summarized as follows:

1. Develop methods to better understand recharge to the Memphis aquifer and to determine the nature of the hydraulic connection between the shallow (alluvial/fluvial) aquifer system and the Memphis aquifer at a specific window location;
2. Provide more reliable information on recharge and the potential for contaminant migration into the Memphis aquifer; and
3. Demonstrate the utility of emerging environmental tracer techniques in age-dating groundwater and for flow-model calibration.

STUDY AREA

The study site selected for the experimental design is the Shelby Farms window site. The Shelby Farms site is located in the eastern sector of Shelby County, between the primary urban center and rapidly developing suburban communities. The site is bordered by the Wolf River to the southwest and is characterized as Wolf River flood plain. Further, the site is characterized as a relatively flat area with levees, drainage ditches, and intermittent streams. Much of the area is still maintained as agricultural lands. The site has undergone extensive hydrogeologic investigation subsequent to the closure of the municipal landfill in 1988 (Bradley, 1988; Bradley, 1991; Parks and Mirecki, 1992). A detailed timeline of these studies was presented by Parks and Mirecki (1992).

aquifer. The goals of the study, as presented in previous progress reports, are summarized as follows:

CHAPTER 2
SHELBY FARMS HYDROGEOLOGY AND CONCEPTUAL MODEL

RESULTS FROM PREVIOUS STUDIES

The Shelby Farms window site was selected for this study on the basis of previous research, a pre-existing data base, and a research infrastructure at the site. A map of the Shelby Farms study area showing the locations of monitoring wells and other topographic features is provided in Figure 2.1. Early studies in the Shelby Farms area included focused investigations on the hydrogeology and potential for contamination of the Memphis aquifer from the Shelby County landfill following identification of a window located just north of the landfill (Bradley, 1988; Bradley, 1991; Parks and Mirecki, 1992). Results from these studies documented (1) thinning or absence of the clay confining unit separating the alluvial (shallow) aquifer from the Memphis aquifer, (2) a depression in the water table in the alluvial aquifer and corresponding mounding of the potentiometric surface of the Memphis aquifer in the area of the window indicating downward leakage of water through the window, (3) confirmation that selected inorganic and organic constituents of leachate from the landfill have migrated northward in the alluvial aquifer and downward through the window into the Memphis aquifer, and (4) that groundwater would require about 50 to 150 years to travel from the landfill to the closest downgradient public-supply well field. Contaminants in the groundwater would not likely persist to reach the well field because of various physical, chemical, and biological processes, including dilution and adsorption. As part of these studies and a follow-up study, tritium (^3H) samples were collected from select wells by the USGS in 1987 and by the Ground Water Institute (GWI) at the University of Memphis in 1999 prior to the onset of this study to evaluate leakage at the Shelby Farms window site. The results of the ^3H analyses are provided in Table 2.1. Since values of ^3H in excess of background (~0.1 to 0.5 tritium unit (TU) where 1 TU equals approximately 3.19 picocuries per liter) indicate the presence of modern water, these data confirm the presence of modern water in the upper part of the Memphis aquifer at the Shelby Farms site.

DATA COLLECTED FROM EXPLORATORY DRILLING

During the week of March 26, 2001, three exploratory test holes were drilled at separate locations at the Shelby Farms study area (1) to refine the western and northwestern extent of the window previously identified in the upper Claiborne confining unit by investigating the presence of clay at these locations, and (2) to select the best location for one of two proposed clusters of observation wells for the "confined" site located downgradient with respect to the window and the prevailing groundwater flow direction in the Memphis aquifer. The locations of the three test holes are shown on Figure 2.2. The test holes, numbered 1 through 3, were drilled to depths of 120, 140, and 130 feet below land surface, respectively, by the U.S. Army Corps of Engineers (USACE), Memphis District using hydraulic (mud) rotary drilling methods. Geologist's logs of the lithologies encountered were recorded and grab samples of the drill cuttings were collected at approximately 10 foot intervals during drilling. In addition to samples of the cuttings, two Shelby tube samples (3 foot long by 3 inch diameter) of the confining unit clay were collected for laboratory analyses from the ~45-48 foot and the ~60-63 foot intervals in Test Hole #3. Upon reaching the total drilling depth in each test hole, natural gamma-ray and electric (single-point resistance

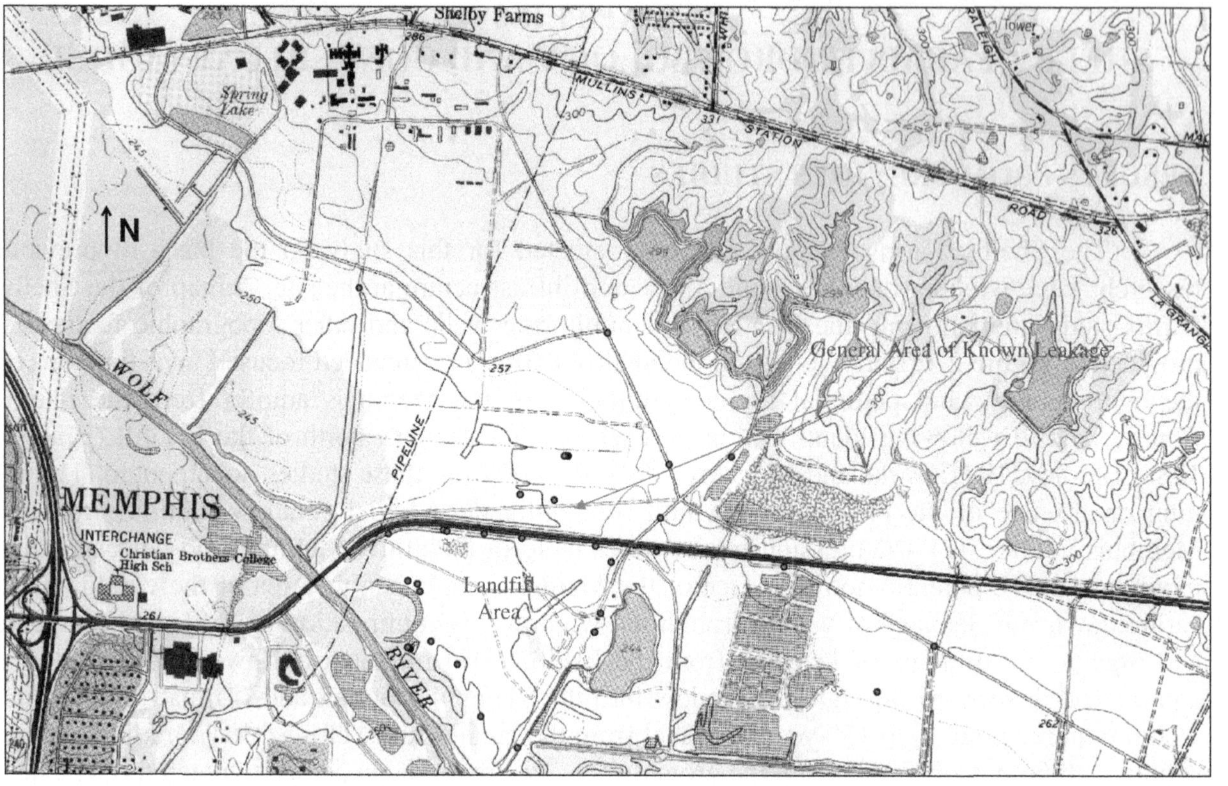

Source: Adapted from USGS 1993.

Figure 2.1 Site map for the Shelby Farms study area

Table 2.1
Tritium data from the Shelby Farms study area

Shelby Farms well ID	USGS local well number	Depth, in feet	Hydrogeologic unit	1999 tritium results (in TU)	1987 tritium results (in TU)
1	Sh:Q-95	35.6	Alluvium	4.07	9.1
30	Sh:Q-128	38.7	Alluvium	—	9.4
31	Sh:Q-129	39	Alluvium	4.61	—
34	Sh:Q-134	38.5	Alluvium	3.75	—
MS-16	Sh:Q-123	83.5	Confining unit	4.55	—
MS-1	Sh:Q-90	170	Memphis aquifer	4.17	—
MS-2	Sh:Q-92	180	Memphis aquifer	4.08	10.6 9.7
MS-3	Sh:Q-125	100	Memphis aquifer	—	4.7
MS-4	SH:Q-126	97.8	Memphis aquifer	—	0.7
MS-9	Sh:Q-148	117.5	Memphis aquifer	2.79	—
MS-11	Sh:Q-150	133.5	Memphis aquifer	8.01	—
MS-12	Sh:Q-151	87.5	Memphis aquifer	4.71	—

Source: Adapted from Bradley (1991) and Gentry (2001).
TU – Tritium unit; — – no sample.

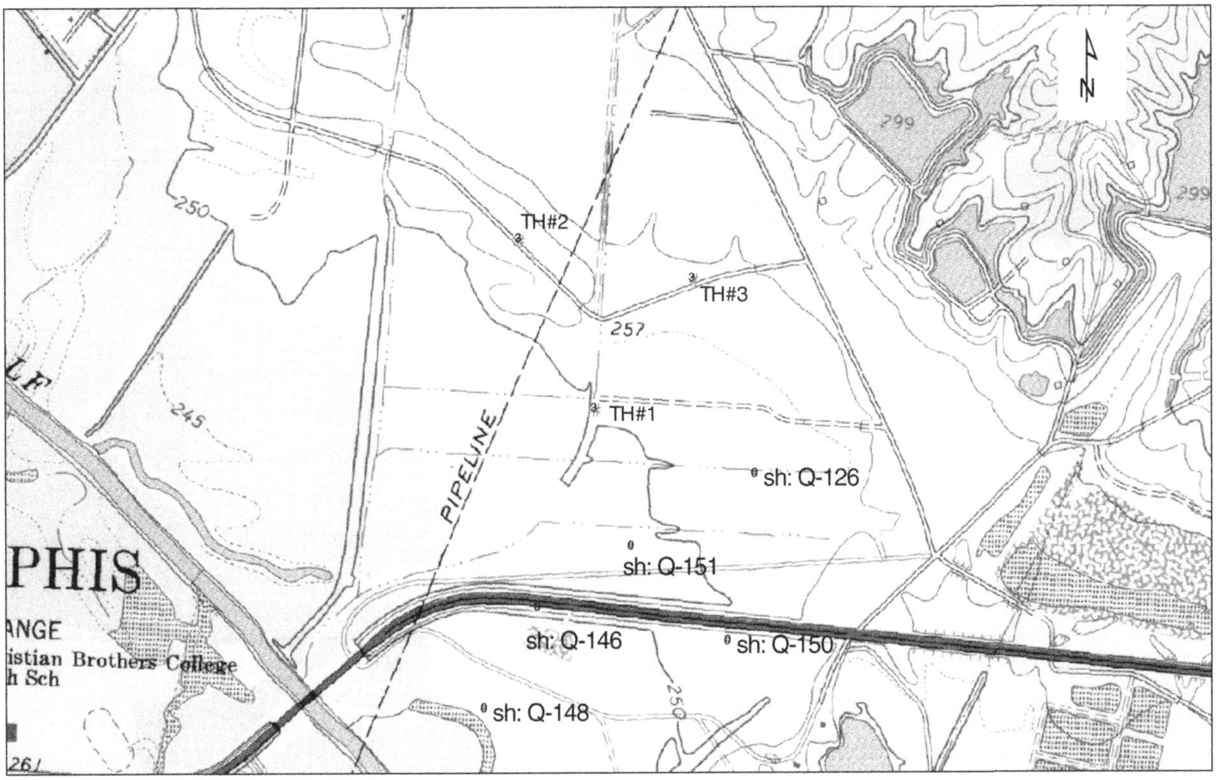

Source: Adapted from USGS 1993.

Figure 2.2 Location of exploratory borings at the Shelby Farms study area in Shelby County, TN

and spontaneous potential) geophysical logs were made in the drilling-mud-filled borings. Following logging, the test holes were abandoned by pumping a Portland cement/powdered bentonite mixture through the drill rods from the bottom of each boring upward to the land surface. The depths, lithologies, and hydrogeologic units encountered during drilling, as identified from the geophysical logs, are provided in Table 2.2.

Geophysical data are presented on Figures 2.3, 2.4 and 2.5 as they relate to responses on the individual geophysical logs. As seen in Table 2.2, clay of the upper Claiborne confining unit was encountered in all three test holes, although it appeared to be of a different character in Test Hole #2, the most northwestern of the three borings (see Figure 2.2). The presence of clay in the three test holes provides improved definition of the western and northwestern extent of the identified window in the confining unit. Following completion of the test holes and evaluation of the data collected from the drilling program, the vicinity of Test Hole #1 was selected as the preferred location for the "confined" well cluster because of the thickness and competency of the confining unit in this area, and because of the downgradient position of the cluster relative to the window and groundwater-flow direction for the Memphis aquifer in the study area (see Figure 2.2). A crosssection through the study area using selected borings is shown in Figure 2.6. The profile demonstrates the geologic evidence for the location of the window site and thinning of the upper Claiborne confining unit.

Table 2.2
Lithology descriptions and hydrogeologic unit designation based on test-hole drilling

Test hole	Depth, feet	Thickness, feet	Lithology	Hydrogeologic unit
#1	0 – 8	8	Silt and clayey silt	Upper alluvium
	8 – 35	27	Sand and gravel	Shallow aquifer
	35 – 82	47	Clay, w/ some silty clay and fine sand	Upper Claiborne confining unit
	82 – 120	*	Very fine to coarse sand	Memphis aquifer
#2	0 – 17	17	Silt and clayey silt	Upper alluvium
	17 – 48	31	Sand and gravel	Shallow aquifer
	48 – 92	44	Silty clay w/ very fine sand	Upper Claiborne confining unit
	92 – 140	*	Very fine to very coarse sand	Memphis aquifer
#3	0 – 11.5	11.5	Silt and clayey silt	Upper alluvium
	11.5 – 43	31.5	Sand and gravel	Shallow aquifer
	43 – 95	52	Clay, w/ some silty clay and fine sand	Upper Claiborne confining unit
	95 – 130	*	Very fine to very coarse sand	Memphis aquifer

* Thickness values not given for Memphis aquifer because test holes penetrated only uppermost part of unit.

ROTASONIC DRILLING AND GEOPHYSICAL LOGGING

In August 2001, a drilling program was conducted that consisted of the installation of nine observation wells at four locations in the Shelby Farms study area by rotasonic methods; field description and collection of core from selected borings for geotechnical and geochemical analysis; and the running of geophysical logs in selected borings at three of the four sites. The locations of the wells are shown on Figure 2.7.

Three of the wells were installed at the "unconfined" (window) location, and together with an existing well installed from a previous study (MS-12), compose a four-well cluster with one well screened in the alluvium of the Wolf River (shallow aquifer) and three wells screened at successively greater depths in the upper part of the Memphis aquifer (including the pre-existing well). Four of the nine wells were installed at the "confined" location, which was relocated about 500 feet southwest of the location of Test Hole #1 to achieve a position more downgradient of the unconfirmed cluster and with fewer drilling equipment access problems than at the original Test Hole #1 location. The confined cluster consists of a well screened each in the alluvial aquifer, confining unit, and two zones of the upper part of the Memphis aquifer. Also, as part of this drilling program, two additional wells were installed and screened in the upper part of the Memphis aquifer at separate locations, each within 20 feet of existing single wells screened in the shallow aquifer (southwestern and northwestern nests shown on Figure 2.7). Data for the wells are provided in Table 2.3.

Rotasonic cores were collected from the deepest borings at the confined and unconfined locations and from selected intervals of the boring at the northwestern well nest. The cores from the southwestern and northwestern well nests were described in detail and sampled at discrete intervals at the site prior to discarding any portion of the cores. The rotasonic cores retained from the borings are stored at the University of Memphis and were used for detailed laboratory determinations,

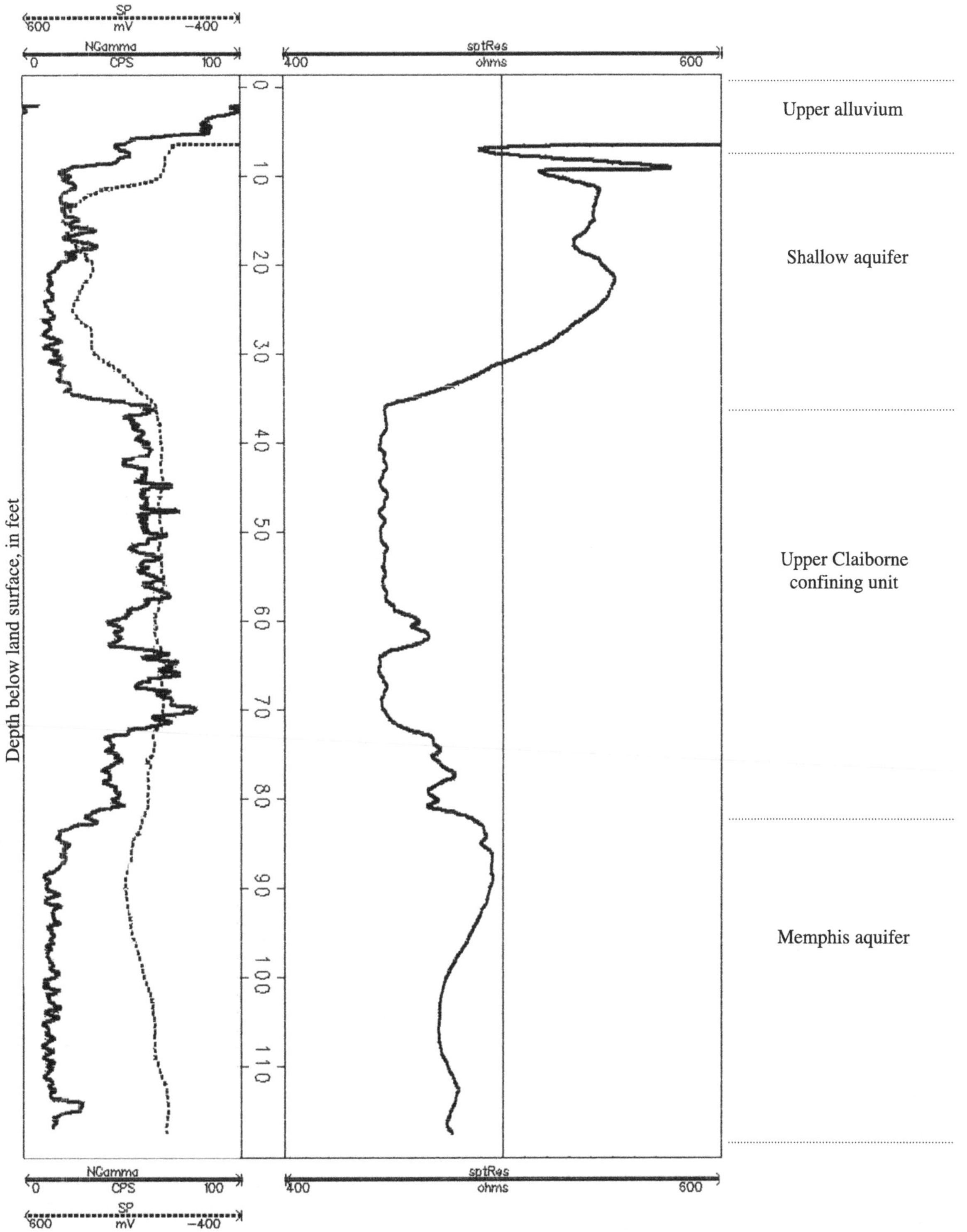

Figure 2.3 Geophysical log and hydrogeologic unit identification for Test Hole #1

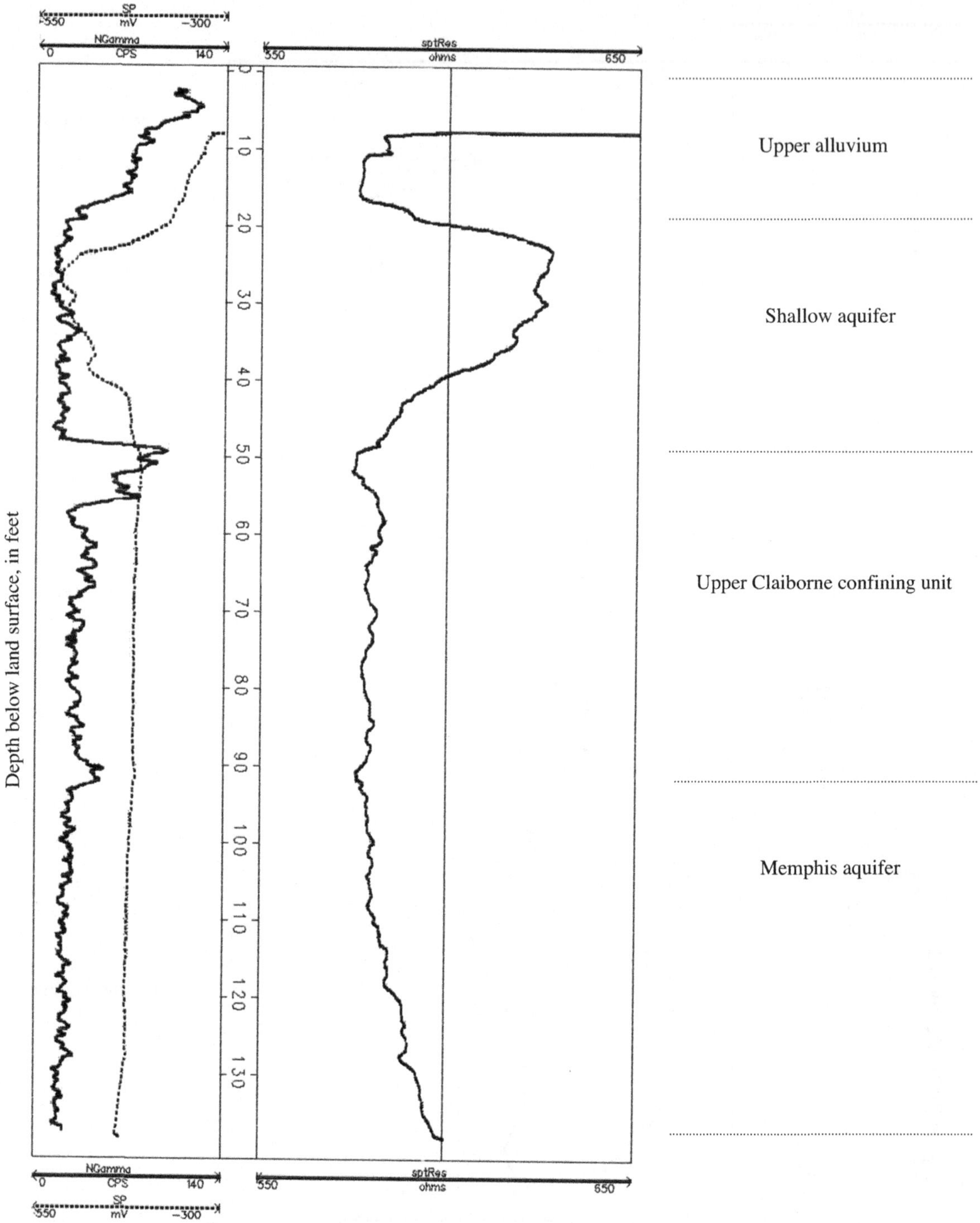

Figure 2.4 Geophysical log and hydrogeologic unit identification for Test Hole #2

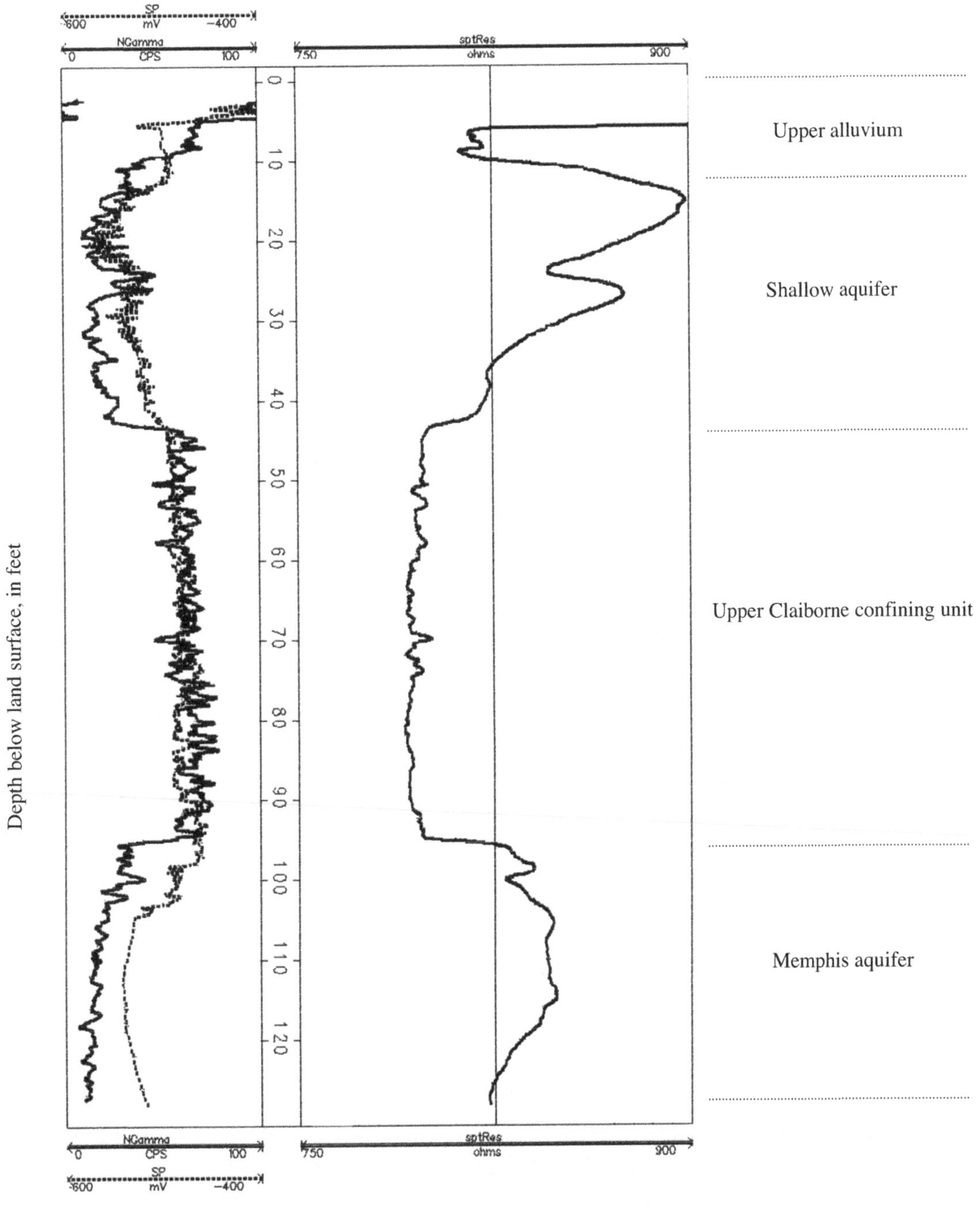

Figure 2.5 Geophysical log and hydrogeologic unit identification for Test Hole #3

including grain-size and organic-carbon analysis. Natural gamma-ray geophysical logs were run through the steel rotasonic casing in the deep boring at the confined location, in the second deepest boring at the unconfined location, and in the boring at the southwestern well nest (see Figure 2.7). Problems with rotasonic drill penetration and core recovery caused the planned deepest well at the unconfined location to be completed shallower than anticipated, but a deeper well was subsequently completed at this location; see the following paragraph on drilling and well-installation specifics for

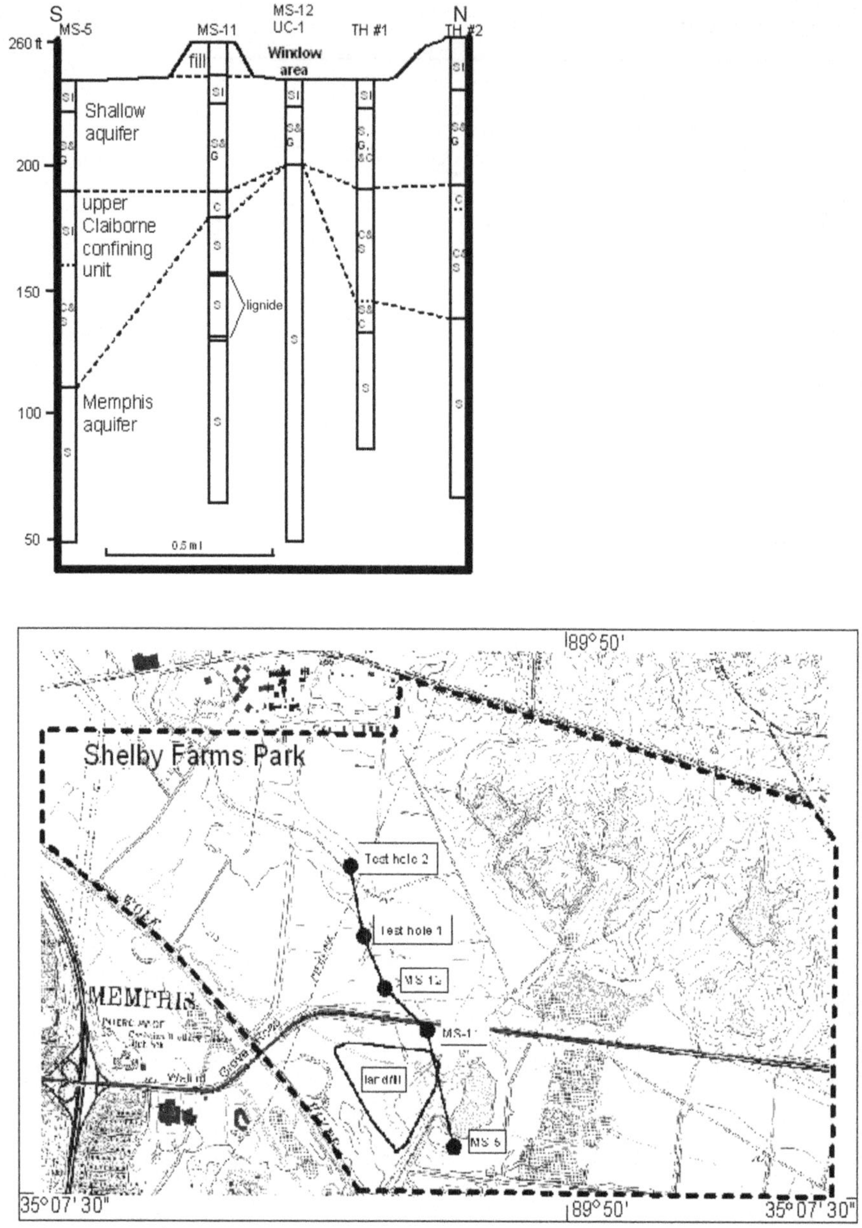

Figure 2.6 Crosssection through Shelby Farms study area and stratigraphic profiles for selected borings (SI – silt, S&G – sand and gravel, C – clay, S – sand)

details. The boring at the southwestern well nest was drilled deeper than the final screened interval of the well because of the total thickness of the confining unit in this boring (aggregate clay thickness of equal to or greater than ~127 feet). In August 2003, electromagnetic (EM) induction logs were run through the PVC casing in the deepest wells at the unconfined and confined cluster sites and the single well installed at the southwestern well nest, and both gamma and EM induction logs were run in the single well installed at the northwestern well nest to complete the suites of geophysical logs for wells installed in the study area during this investigation.

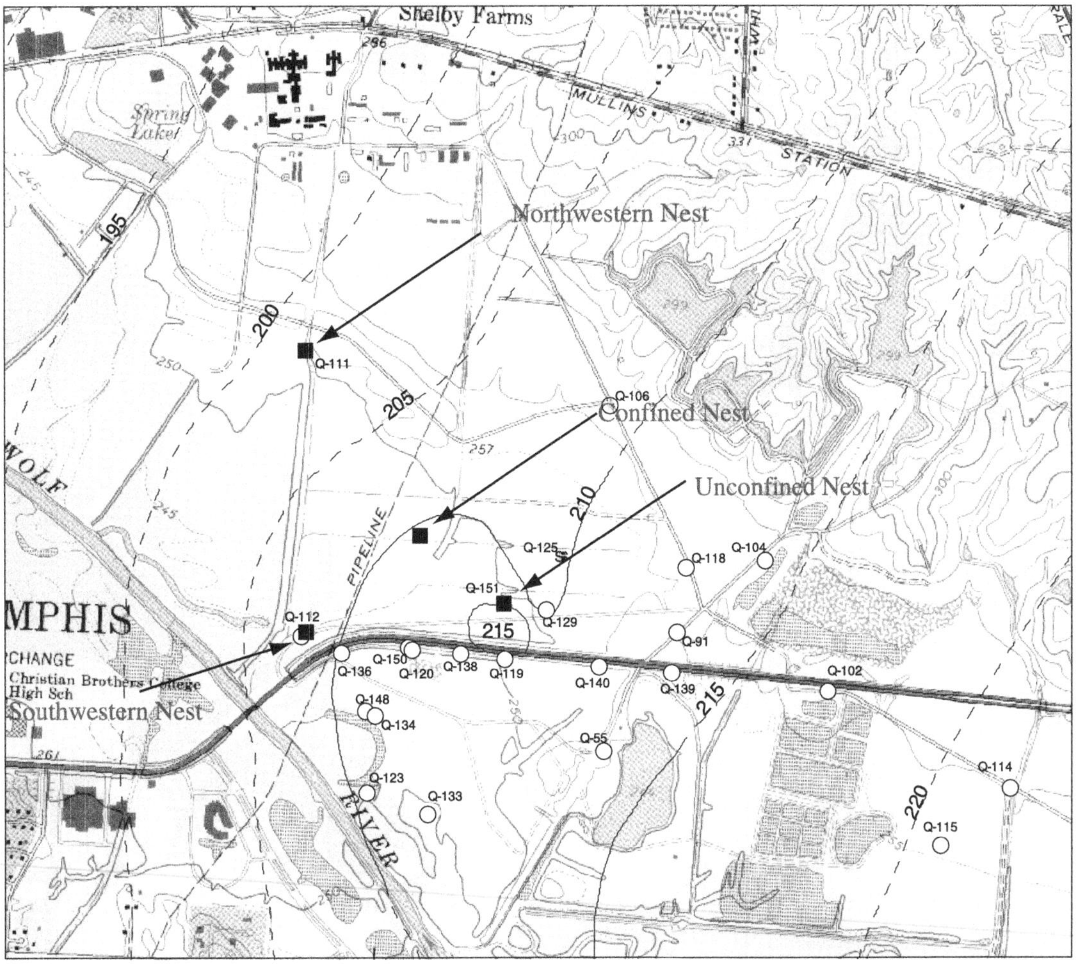

1:19668

■ Well Nests

○ Observation Wells

/\/ Potentiometric Surface, 5-foot contour interval

N

Source: Adapted from USGS (1993) and Parks and Mirecki (1992).

Figure 2.7 Potentiometric surface of the Memphis aquifer, October 1989, and well-nest layout for Shelby Farms study area

During drilling and installation of the wells at the two cluster locations, a variety of problems were encountered that, for the most part, were overcome by adjusting drilling or well-construction specifications in the field. Because the rotasonic drilling method has had only limited use in drilling into the Memphis aquifer in the Memphis area (at the time of well installations for this project, only one well was known to have been installed into the Memphis aquifer), and had not been used for drilling in the Shelby Farms study area, the very fine sand present in the upper part of the Memphis aquifer at the well-cluster locations caused poor drill penetration, poor core recovery, and/or problems with sand heaving into the lower part of the rotasonic casing during drilling of the deeper borings at these two sites. At the unconfined location, the first well cluster to

Table 2.3
Observation well data

Project well number*	USGS local well number†	Date drilled	Latitude—longitude‡ (degrees, minutes, and seconds)	Land surface altitude, in feet§ (AMSL)**	Top of casing (TOC) altitude, in feet§ (AMSL)**	(Boring Diameter)††/ [core diameter], in inches	Boring depth, in feet BLS‡‡	Well depth, in feet (below TOC)	Casing and screen diameter (in inches)/ material	Screen interval, in feet BLS/unit screened
MS-13 (UC-1)	Sh:Q-171	8/11/01	350811 0895044	247.92	249.77	(8) / [6 and 4]	~145	118.27	2/Schedule 40 PVC§§	~110-115/ Memphis aquifer
MS-14 (UC-2)	Sh:Q-172	8/14/01	350811 0895044	247.92	249.99	8 and (6)/[4]	~155	151.91	2/Schedule 40 PVC§§	~144-149/ Memphis aquifer
44 (UC-3)	Sh:Q-173	8/15/01	350811 0895044	247.92	249.79	(6)/ [4]	~55	51.66	2/Schedule 40 PVC§§	~45-50/ alluvial aquifer
MS-15 (C-1)	Sh:Q-174	8/22/01	350818 0895056	246.37	248.37	6 and (8)/[4]	~158	159.48	4/Schedule 40 PVC§§	~137-157/ Memphis aquifer
MS-16 (C-2)	Sh:Q-175	8/23/01	350818 0895056	246.37	248.30	(6)/[4]	~100	99.50	2/Schedule 40 PVC§§	~93-98/ Memphis aquifer
45 (C-3)	Sh:Q-176	8/26/01	350818 0895056	246.37	248.15	8/[(6)]	~61.5	63.36	2/Schedule 40 PVC§§	~58-61/ confining unit
46 (C-4)	Sh:Q-177	8/26/01	350818 0895056	246.37	248.09	(6)/[4]	~36.5	38.52	2/Schedule 40 PVC§§	~31.5-36.5/ alluvial aquifer
MS-17 (19-D)	Sh:Q-178	8/29/01	350807 0895111	246.98	249.01	(6)/[4]	~210	144.39	2/Schedule 40 PVC§§	~132-142/ Memphis aquifer
MS-18 (18-D)	Sh:Q-179	8/29/01	350838 0895112	260.81	262.64	(6)/[4]	~166.5	168.16	2/Schedule 40 PVC§§	~155-165/ Memphis aquifer

(continued)

Table 2.3 (Continued)

Project well number*	Screen slot size, in inches	Sand pack interval, in feet BLS	Sand pack material	Seal material	Seal interval, in feet BLS	Boring backfill material	Backfill interval, in feet (0 = land surface)	Gallons of water used in drilling***	Drilling mud added
MS-13 (UC-1)	0.010	~108-116	10-30 sieve silica sand	3/8" bentonite chips	~106-108	Powdered bentonite/ water	0-~106	~2,000	No
MS-14 (UC-2)	0.010	~141-150	10-30 sieve silica sand	3/8" bentonite chips	~138-141	Powdered bentonite/ water	0-~135	unrecorded	Yes
44 (UC-3)	0.010	~43-51	10-30 sieve silica sand	3/8" bentonite chips	~41-43	Powdered bentonite/ water	0-~41	unrecorded	No
MS-15 (C-1)	0.010	~134-157.5	10-30 sieve silica sand	3/8" bentonite chips	~131-134	Powdered bentonite/ water	0-~131	unrecorded	Yes (below 85')
MS-16 (C-2)	0.010	~90.5-97.5	10-30 sieve silica sand	3/8" bentonite chips	~88-90.5	Powdered bentonite/ water	0-~88	unrecorded	No
45 (C-3)	0.010	~56-61.5	10-30 sieve silica sand	3/8" bentonite chips	~46-56	Powdered bentonite/ water	0-~46	unrecorded	No
46 (C-4)	0.010	~29-36.5	10-30 sieve silica sand	3/8" bentonite chips	~24-29	Powdered bentonite/ water	0-~24	unrecorded	No
MS-17 (19-D)	0.010	~130-143	10-30 sieve silica sand	3/8" bentonite chips	~125-130	Powdered bentonite/ water	0-~125	unrecorded	No
MS-18 (18-D)	0.010	~154-166.5	10-30 sieve silica sand	3/8" bentonite chips	~148-154	Powdered bentonite/ water	0-~148	unrecorded	No

(continued)

Table 2.3 (Continued)

Project well number*	Depth to water, in feet BLS	Water-level altitude, in feet AMSL	Measuring point†††, in feet	Date of water-level measurement	Type of development	Gallons of water removed during development	Remarks
MS-13 (UC-1)	42.20	205.87	~1.85	11/6/01	Surged/pumped	~190 (partially developed)	Entire core to 145' saved in 22 boxes; 6" core to 0'-115', 4" core 115'-145'; hole collapsed from ~116'-145' before making well; ~6"-long well point (sump) at bottom of screen; 4' × 6" round steel well protector and 4, 5' × 3" bumper posts installed; purge water very turbid and strong CH_4/H_2S smell at start of development; cleared to "milky"/clear color at end of development; gamma-ray log 2.75'-128.5'
MS-14 (UC-2)	44.20	203.89	~2.07	11/6/01	Surged/pumped	~160 (partially developed)	Core from 135'-145' added to box #22 (137'-145' interval) from MS-13; core from 145'-155' in box #23; 4" core 0'-155'; hole collapsed to ~150' before making well; may have formation collapse above seal from ~135'-138'; ~6" well point (sump) at bottom of screen; 4' × 6" round steel well protector and 4, 5' × 3" bumper posts installed; purge water very turbid and moderate CH_4/H_2S smell at start of development; water very clear at end of development; preliminary aquifer test (step) conducted 4/18/02
44 (UC-3)	41.51	208.28	~1.87	11/5/01	Surged/pumped	~140 (partially developed)	No core saved; hole backfilled with bentonite chips from ~51'-55' before making well; ~1"-long flat well cap (sump) at bottom of screen; 4' × 6" round steel well protector and 4, 5' × 3" bumper posts installed; purge water very turbid and strong CH_4/H_2S smell at start of development; cleared to "milky"/clear color at end of development

(continued)

Table 2.3 (Continued)

Project well number[*]	Depth to water, in feet BLS	Water-level altitude, in feet AMSL	Measuring point[†††], in feet	Date of water-level measurement	Type of development	Gallons of water removed during development	Remarks
MS-15 (C-1)	44.11	202.26	~2.00	11/7/01	Surged/pumped and pumped	~185 (partially developed)	Entire core to 155' saved in 27 boxes; 4" core 0'-155'; 8" casing overdrilled 6" casing to ~158' before making well; ~6" flat well cap (sump) at bottom of screen; 4' × 8" round steel well protector and 4, 5' × 3" bumper posts installed; purge water very turbid and strong CH_4/H_2S smell at start of development; 4" pump installed temporarily in well to purge intermittently after recovery; well pumps/recovers only at ~0.3 gpm and discharge contains very fine sand; gamma-ray log 2.75'-155'; preliminary aquifer tests conducted 4/16 and 4/18/02
MS-16 (C-2)	44.00	202.30	~1.93	11/7/01	Surged/pumped	~350 (partially developed)	No core saved; hole collapsed from ~97.5'-100' before making well; ~6"-long well point (sump) at bottom of screen; 4' × 6" round steel well protector and 4, 5' × 3" bumper posts installed; purge water very turbid and strong CH_4/H_2S smell at start of development; discharge still cloudy and slight CH_4/H_2S smell at end of development; discharge contains very fine sand
45 (C-3)	49.94	196.41	~1.78	11/7/01	Not developed	N/A	No core saved; 8" casing and 6" core barrel used to 42' (~7' into clay confining unit), water in 8" casing then removed w/ pump and bailer, then 6" core barrel only used from ~40'- 61.5' and drilled dry; top of seal ~10' below top of clay; ~6"-long well point (sump) at bottom of screen; 4' × 6" round steel well protector and 4, 5' × 3" bumper posts installed; well not developed
46 (C-4)	33.98	212.31	~1.80	11/7/01	Surged/pumped	~180 (partially developed)	No core saved; ~6"-long well point (sump) at bottom of screen; purge water very turbid and strong CH_4/H_2S smell at start of development; cleared to light brown color at end of development

(continued)

Table 2.3 (Continued)

Project well number*	Depth to water, in feet BLS	Water-level altitude, in feet AMSL	Measuring point†††, in feet	Date of water-level measurement	Type of development	Gallons of water removed during development	Remarks
MS-17 (19-D)	46.62	200.39	~2.03	11/6/01	Surged/pumped	~120	No core saved; hole backfilled with bentonite chips from ~144'-203' after drilling; formation collapsed from ~143'-144' before well set; ~1"-long flat well cap (sump) at bottom of screen; drillers pumped ~60 gallons water from well after completing; 4' × 6" round steel well protector and 4, 5' × 3" bumper posts installed; water very turbid and strong CH_4/H_2S smell at start of development water very clear at end of development; gamma-ray log 2.75'-200'
MS-18 (18-D)	61.14	197.87	~1.83	11/8/01	Surged/pumped	~200	Core saved from ~55'-155'; ~6"-long well point (sump) at bottom of screen; 4' × 6" round steel well protector and 4, 5' × 3" bumper posts installed; purge water very turbid and strong CH_4/H_2S smell at start of development; discharge still cloudy at end of development; discharge contains very fine sand

* Well name/number convention used for wells installed for previous studies at Shelby County Landfill (SCLF); Memphis Sand (MS) wells installed for this study assigned numbers in sequence with those used for SCLF studies; wells 44-46 are alluvial wells installed for this study also using numbers in sequence with those assigned to shallow wells installed for SCLF studies; wells MS-18 and MS-17 are MS wells paired with alluvial wells 18 and 19 from previous SCLF studies and are assigned "D" for deepest well of pair; numbers in parentheses are other project well numbers used during study; UC – unconfined, C - confined

† USGS local well number for Shelby County

‡ From Global Positioning Satellite readings; NAD 83 datum

§ From survey data

** Above mean sea level

†† Diameter of rotasonic casings; parentheses indicate diameter of casing in which well was constructed

‡‡ Below land surface

§§ PVC – Polyvinyl chloride

*** Estimate of water used to drill hole; does not include water used to mix bentonite grout backfill

††† Measuring point is approximate height of TOC above land surface

CH_4 – methane

H_2S – hydrogen sulfide

~, approximate

N/A – not applicable

', feet; ", inches

gpm, gallons per minute

be drilled, the program called for drilling of the deepest and largest diameter boring first to run geophysical logs through the entire interval of interest, followed by installation of a 4-inch-diameter well that would serve as the production well for planned aquifer tests. Results from the first boring/well then would be used to finalize projected completions depths for the shallower wells. After several unsuccessful attempts to alter the drilling procedures to complete the deep well to its projected depth, the initial boring at this location was terminated at a shallower depth and a smaller, 2-inch-diameter well was completed in the boring. Subsequent to completion of the first well, a second boring adjacent to the first well was completed to the intended deepest depth for this cluster; however, use of smaller diameter drill casing and core barrels were necessary to achieve the deepest depth intended for this cluster, so only a 2-inch-diameter well could be installed in this boring.

Drilling plans at the confined location also called for installation and logging of the deepest and larger diameter boring first, with shallower well completion depths to be based on results from the deep boring/well. Similar problems were encountered with the two deepest wells at the confined location; however, lessons learned while drilling at the unconfined location allowed for the successful yet still difficult installation of a deep, 4-inch-diameter well and slightly shallower, 2-inch-diameter well in the Memphis aquifer, as planned at this location. Additionally, more water was used (water used for drilling was obtained from Fort Pillow aquifer wells in the MLG&W Shaw Well Field located about 10 miles east of the study area) than anticipated, and the addition of drilling mud was needed to drill the two deepest wells at both cluster locations. Use of drilling mud in the borings was not a part of the original drilling program. Use of mud and additional water during drilling of these wells was not problematic for subsequent work at the unconfined nest because well development and aquifer hydraulic tests conducted prior to collection of samples for water-quality analysis removed most of the fluids introduced during drilling, ensuring the integrity of geochemical and environmental isotope samples derived from these wells. At the confined nest, however, use of drilling mud and probable plugging of the Memphis aquifer adjacent to the screen likely is the main reason that the deepest well at this site does not yield much water, even after considerable development and pumping. Boring and well-construction information are provided in Table 2.3.

The wells were completed at land surface with 6-inch-thick by 4-feet-square well pads. Three-inch-diameter by 5-foot-long steel bumper posts were set in the concrete pads at each corner with the tops about 3 feet above land surface. All well casings were cut off about 2 feet above land surface and either 8-inch-diameter (for the 4 inch well) or 6-inch-diameter (for the 2-inch wells) by 4-foot-long steel well protectors were set in the concrete over the well casings to depths of about 1 foot below land surface. Brass tablets were installed in the concrete pads for all wells in order to record well number and altitude information. In September 2001, a local firm was contracted by the GWI to survey the new wells and provide altitudes of survey markers placed at each well. Subsequently, GWI staff surveyed land surface and top-of-casing for each well. Altitude data for the wells are provided in Table 2.3.

During November 2001, the nine wells were partially developed to clear the casings and sand packs of residual drilling fluids. Development consisted of either surge/pumping through the screened intervals, pumping with a submersible pump, bailing, or a combination of these methods. Information on well development is provided in Table 2.3.

Additional geophysical logging of select wells was conducted by the U.S. Geological Survey (USGS) to complete the well-log data set and for comparison with existing log data. In August 2003, EM induction logs were made in well MS-18 (had not been logged previously), and

EM induction logs were made in wells MS-15, MS-17, and MS-14 to supplement the existing gamma-ray logs obtained previously in the test holes for these wells. The corresponding logs and overall stratigraphy for the unconfined nest, confined nest, northwestern nest, and southwestern nest are shown in Figures 2.8, 2.9, 2.10 and 2.11, respectively.

WELL HYDRAULIC TESTING

Hydraulic testing of selected wells was conducted during April through June of 2002. The field data collection was performed using a combination of Solinst Inc. Levelogger® and In-Situ Inc. miniTROLL® pressure transducers for slug tests, and an In-Situ Inc. Hermit® datalogger and PXD-261® for pumping tests. A Solinst Inc. level, temperature, and conductivity (LTC) Levelogger® was utilized for a preliminary multiple-well aquifer test at the confined site well cluster. The data from this test, however, were analyzed as a single-well test because of the low yield of the pumped well (Sh:Q-174) and no discernable drawdown measured in any of the observation wells instrumented at this site. Based on these results, the project team decided not to perform a full-scale aquifer test at either well nest before sampling for environmental tracers because inducing substantial stress on the aquifer system could affect ambient flow paths, and thus, tracer-calculated age dates of samples collected from the wells.

Slug tests were conducted on all individual wells installed for this study with the exception of wells (MS-15) Sh:Q-174, (45) Sh:Q-176 and (46) Sh:Q-177, and in well (MS-12) Sh:Q-151. Well MS-12 is an existing well installed in 1989 for a previous study (Parks and Mirecki, 1992) at the location designated as the unconfined well site for this study and marking the approximate center of the window. As mentioned in the previous section, the project team did not want to disturb the slow water-level recovery in well 45 after installation, so no slug tests were conducted in this well. Recovery data for the well, derived from drawdowns induced prior to sampling, were analyzed, however, to estimate hydraulic conductivity of the confining unit at this location. The water level in well 46, which is screened in the alluvial aquifer, was not high enough in the screen during the study to allow a pressure transducer and slug in the well at the same time without damaging the transducer. The project team decided not to conduct any slug tests that would have involved the introduction of water to any of the wells that may interfere with future environmental-tracer analyses, so a 10-foot-long section of Schedule 40 PVC pipe was filled with sand and used as a slug for all tests. Both falling-head and rising-head slug tests were conducted on those wells tested, with water levels in most of the wells recovering quickly enough to cause the tests to be of only a few minutes in duration. In a few instances, however, recoveries were slow enough that the water level in a particular well may not have fully recovered before a subsequent test was started. In these cases, incomplete water-level recoveries caused problems in analyzing the data.

All hydraulic data analyses were conducted using Microsoft Excel spreadsheet macros developed by the U.S. Geological Survey (Halford and Kuniansky, 2002). The spreadsheet macros utilized for the data analyses included the Bouwer and Rice method for the slug tests (Bouwer and Rice, 1976), and the Cooper-Jacob straight-line (Cooper and Jacob, 1946) and Step-Drawdown methods (Halford and Kuniansky, 2002) for single-well pumping tests. A storage coefficient of 4×10^{-4} was assumed for the Memphis aquifer for the single-well pumping test analyses. The results of the hydraulic tests are provided in Table 2.4.

The grain-size data from discrete core intervals, which are discussed in following sections, were used to calculate hydraulic conductivity values for the unconfined and confined sites. The

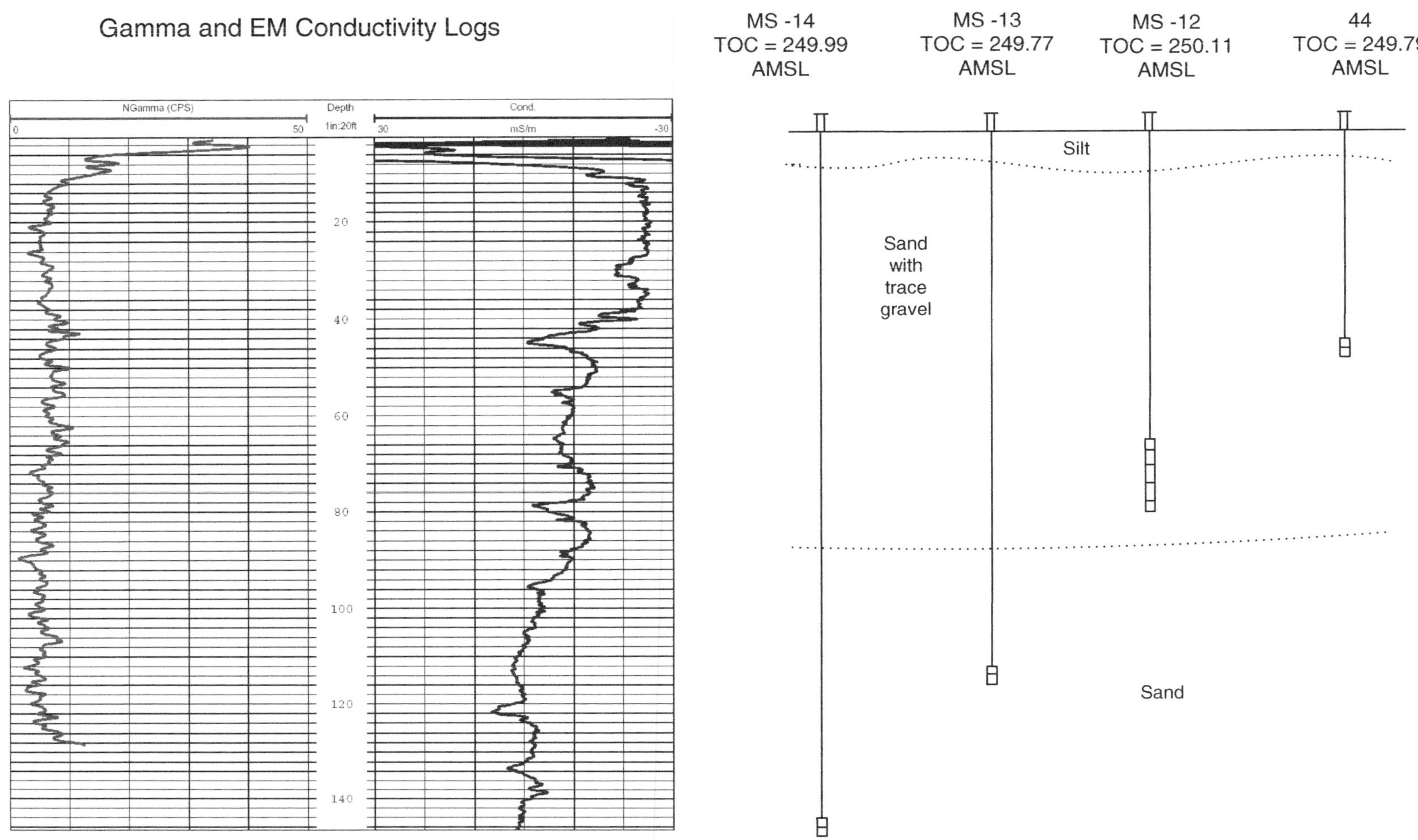

Figure 2.8 Geophysical logs acquired from unconfined nest in August 2001 and 2003 with associated stratigraphic interpretation

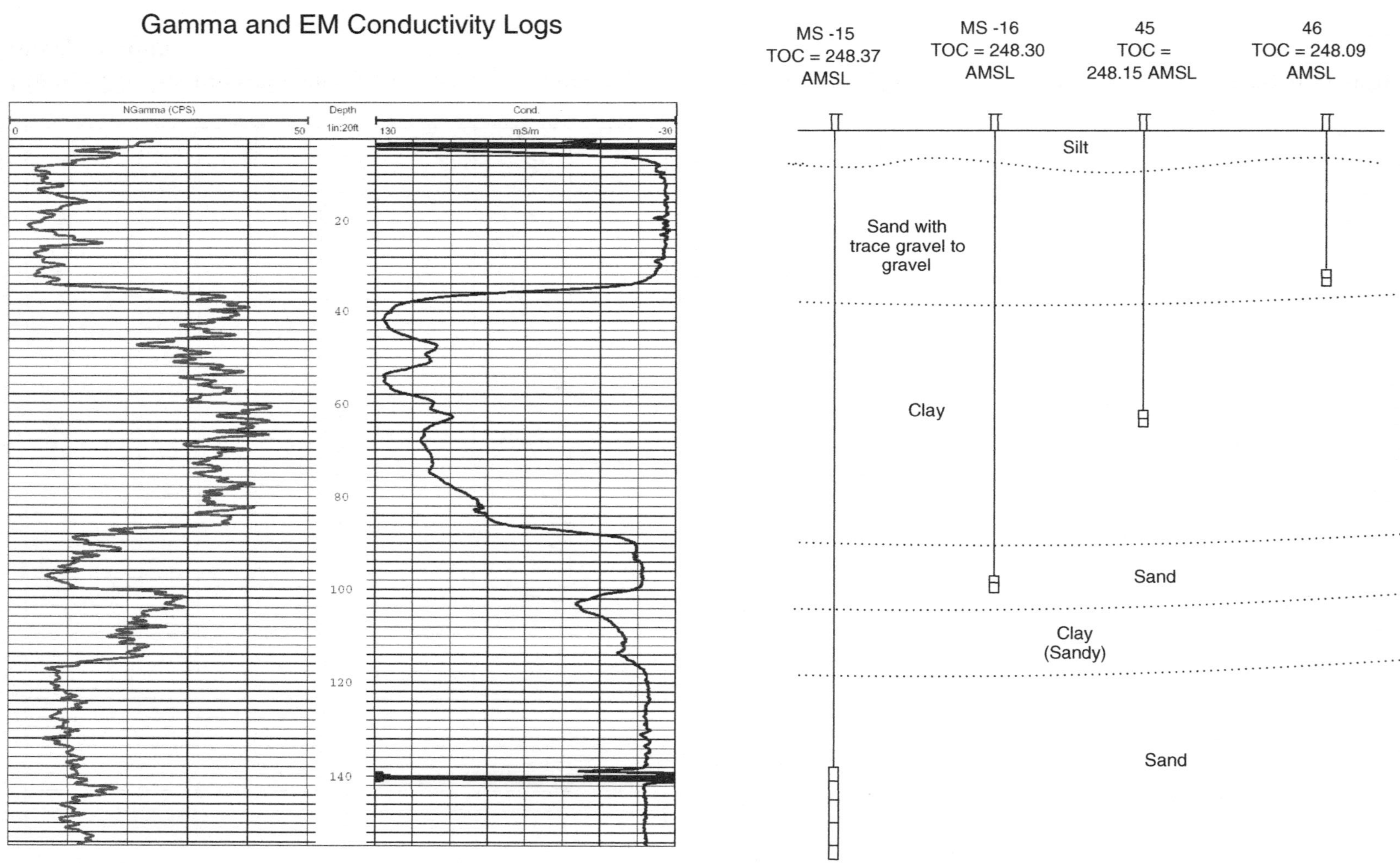

Figure 2.9 Geophysical logs acquired from confined nest in August 2001 and 2003 with associated stratigraphic interpretation

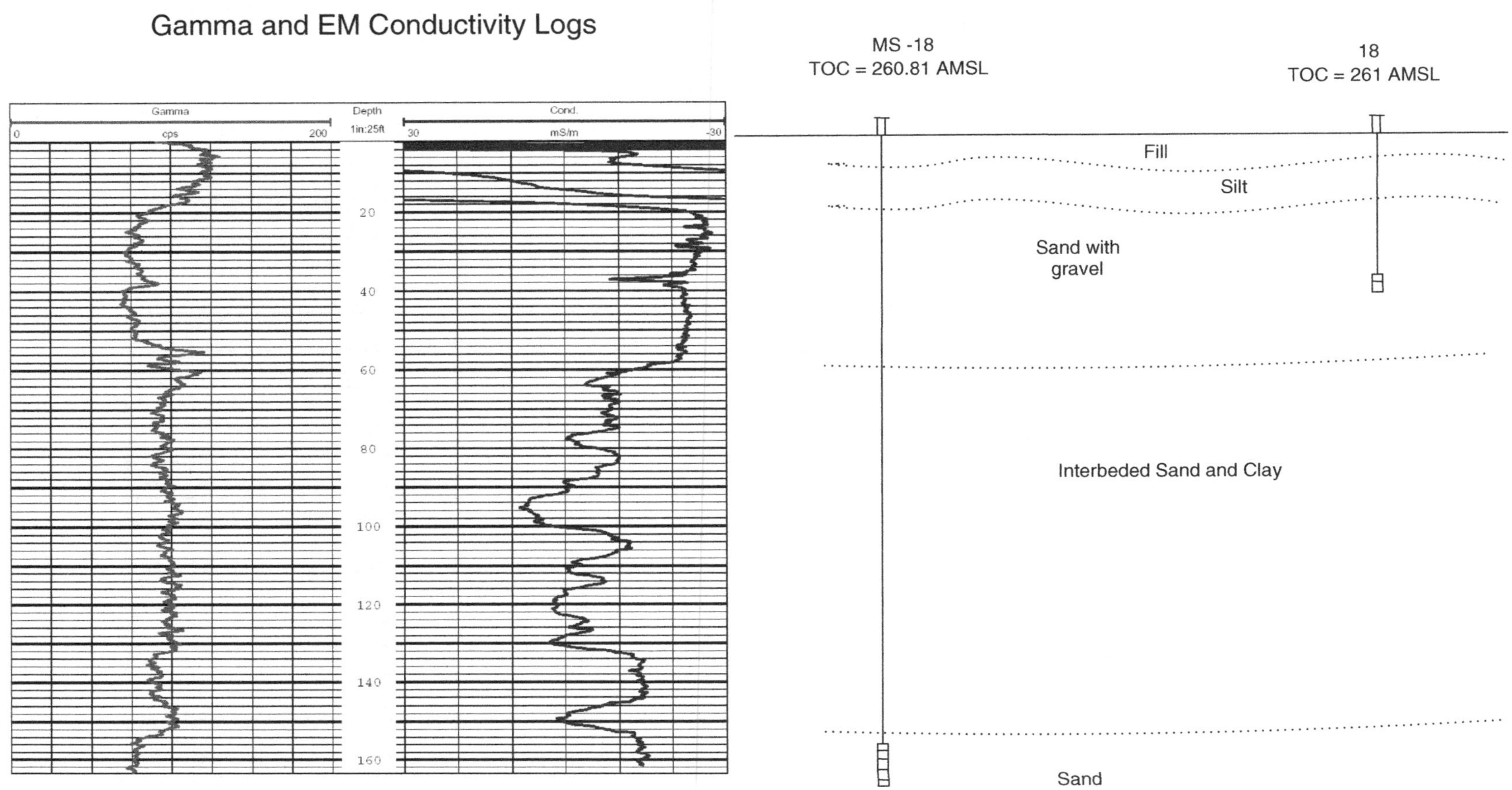

Figure 2.10 Geophysical logs acquired from northwestern nest in August 2003 with associated stratigraphic interpretation

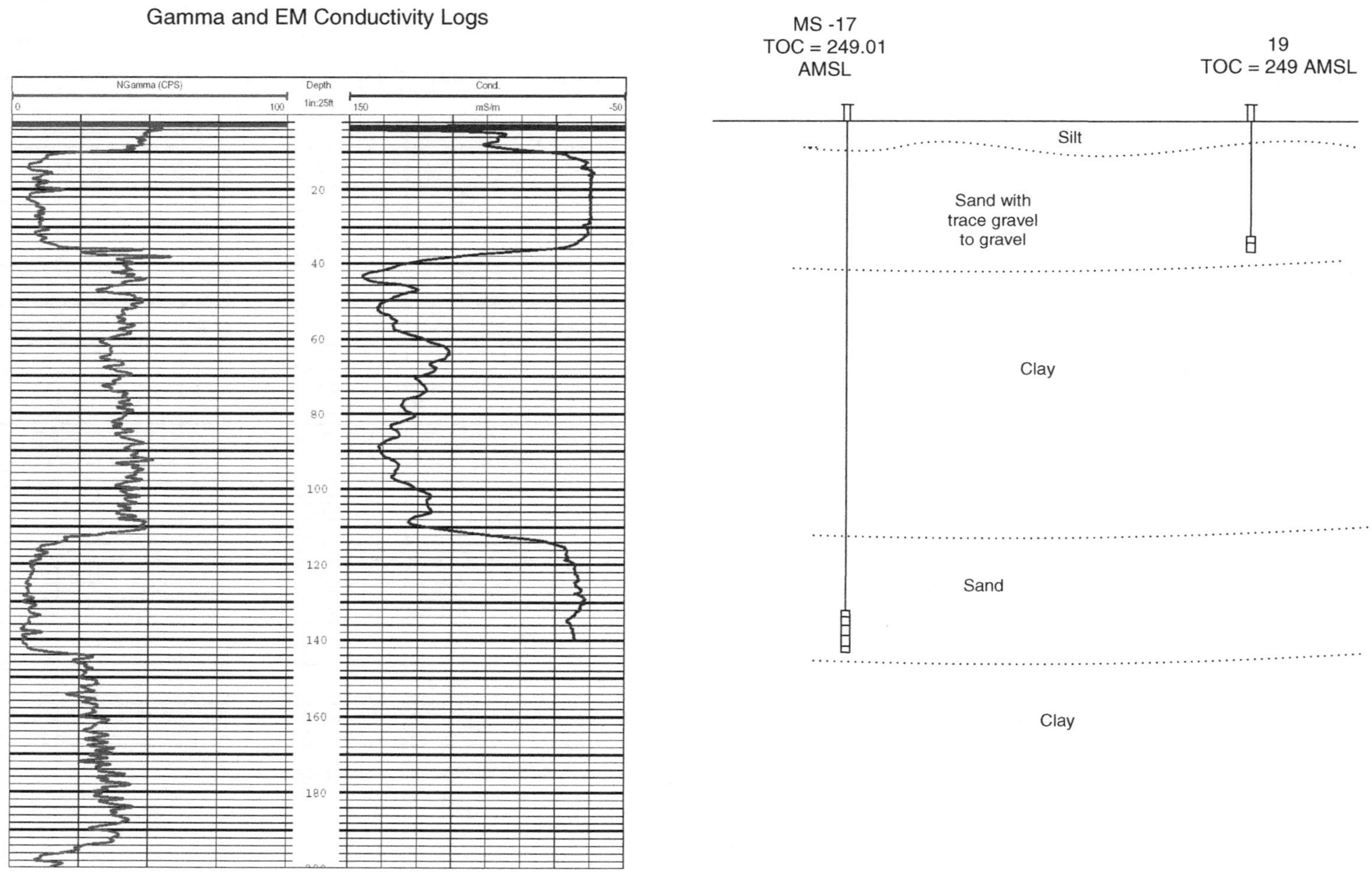

Figure 2.11 Geophysical logs acquired from southwestern nest in August 2001 and 2003 with associated stratigraphic interpretation

Table 2.4
Summary of hydraulic test data

Well ID	Date(s) of test(s)	Test type/number of tests (FH-falling head, RH-rising head)	Analysis method(s)	Screened unit	Range of hydraulic conductivity (ft/day)	Best hydraulic conductivity value (ft/day) and remarks
				Unconfined Nest		
MS-13 (UC-1)	5/17/02	Slug/3 FH, 1 RH	Bouwer and Rice	Memphis Aquifer	10/20	10
MS-14 (UC-2)	4/18/02	Single well	Step Drawdown	Memphis Aquifer	1	1
44 (UC-3)	6/5/02	Slug/1 FH	Bouwer and Rice	Alluvial Aquifer	100	100
MS-12	5/17/02	Slug/1 FH	Bouwer and Rice	Memphis Aquifer	0.5	0.5
				Confined Nest		
MS-15 (C-1)	4/15-16, 18/02	Single well (x2)	Cooper-Jacob	Memphis aquifer	0.002-0.003(DD)-0.0003-0.00009 (REC)	None; well yield is significantly less than that characteristic of the Memphis aquifer and well has not responded to development; residual effects of rotasonic drilling are suspected to be cause for the low yield
MS-16 (C-2)	5/16/02	Slug/2 FH, 1 RH	Bouwer and Rice	Memphis aquifer	6-7/6	6
MS-16 (C-2)	5/16/02	Single well (×2)	Cooper-Jacob (×2)/Step Drawdown (×1)	Memphis aquifer	0.2/0.4	0.2; hydraulic conductivity values determined by single-well pump test are lower than expected for Memphis aquifer; values determined by slug test of well C-2 probably are more representative
45 (C-3)	8/20/03*	Slug	Bouwer and Rice	Confining unit	0.00002	0.00002
45 (C-3)	12/4/03†	Single well	Cooper-Jacob	Confining unit	0.00005	0.00005
				Southwestern Nest		
MS-17 (19-D)	6/6/02	Slug/3 FH, 3 RH	Bouwer and Rice	Memphis aquifer	80-100/30	100; values determined from rising head tests should be equal to or greater than falling head tests; discrepancies indicated possible incomplete well development
				Northwestern Nest		
MS-18 (18-D)	6/5/02	Slug/2 FH, 2 RH	Bouwer and Rice	Memphis aquifer	1-2/1-2	2

Note: DD – drawdown; REC – recovery
* Recovery of well after initial bail-down on 8/20/03 analyzed
† Recovery of well after sampling on 12/4/03 analyzed

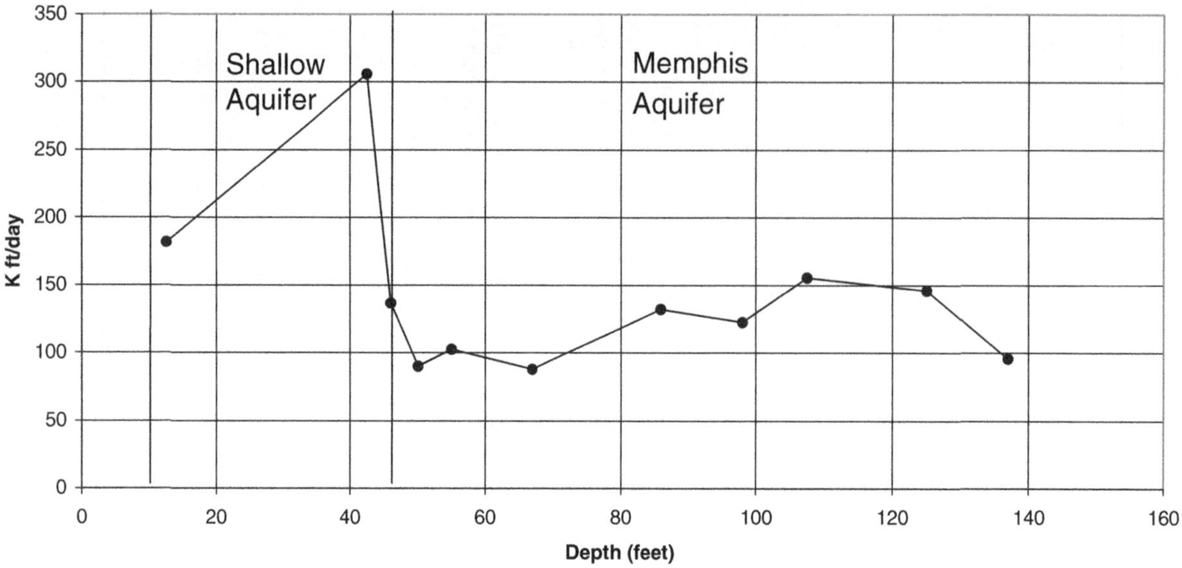

Figure 2.12 Plot of hydraulic conductivity (K) versus depth at the confined site (borehole MS-13 and MS-14). Hydraulic conductivity was calculated using a modified Hazen method of Kasenow, 1997.

results are shown in Figures 2.12 and 2.13. Maximum hydraulic conductivities in the shallow aquifer are approximately 100 ft/day, whereas the hydraulic conductivity values for the Memphis aquifer cluster around 40 ft/day. These values are approximately 30 to 40 ft/day higher than those obtained from hydraulic tests conducted during summer 2002. The grain-size data from the Memphis aquifer do not indicate that any major differences in conductivity exist between the confined and unconfined sites.

WOLF RIVER STREAMFLOW MEASUREMENTS

The U.S. Geological Survey measured stream flow at three locations along the Wolf River in July 2002 to determine the discharge along the section of the River adjacent to the study site. The measurements were made for the purpose of this study to determine the interaction of the Wolf River with the groundwater system at the Shelby Farms study area. Results of the measurements are provided in Table 2.5. Given the accuracy of the measurements, very little flow difference was seen between the three locations along the Wolf River, essentially indicating no gain in flow from Germantown Road to Covington Pike, and potentially demonstrating that some quantity of flow in the Wolf River was being lost to the shallow groundwater system adjacent to the study area.

WATER-LEVEL DATA AND POTENTIOMETRIC MAPS

Water levels in all existing and accessible observation and production wells in the vicinity of the Shelby Farms window site were measured on three occasions over the project period (June 12 to 13, 2002; October 24, 2002; May 27 to 28, 2003). The data were used to prepare potentiometric surface maps for the alluvial aquifer and the Memphis aquifer for late spring/early

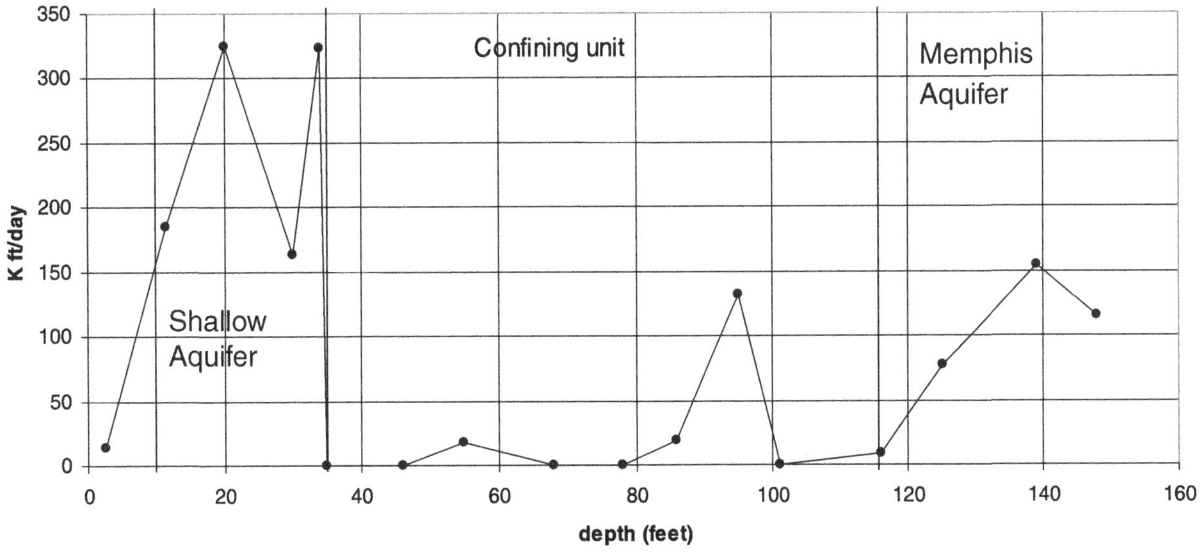

Figure 2.13 Plot of hydraulic conductivity (K) versus depth at the confined site (borehole MS-15). Hydraulic conductivity was calculated using a modified Hazen method of Kasenow, 1997.

Table 2.5
Streamflow measurements of the Wolf River July 2, 2002

Location (bridge crossing)	Flow* (cubic feet second)
Germantown Road	295
Walnut Grove Road	293
Covington Pike	298

* Flow measurements are considered to be accurate within ±5%.

summer (relative seasonal low water level) conditions. The results for June and October 2002 are presented in Figure 2.14 and 2.15.

WATER-LEVEL MONITORING

In February 2003, water-level transducers were deployed in five wells at the Shelby Farms study site. Transducers were installed in the two wells at the southwestern nest [19 and MS-17 (19-D)], two of four wells (deepest and shallowest) at the unconfined nest [MS-14 (UC-2) and 44 (UC-3)], and the shallower of the two wells at the northwestern nest [18]. The objectives of this effort were to collect water-level data from the alluvial and Memphis aquifers in and near the window, and evaluate responses in water levels to changes in stage on the Wolf River; in particular, to high-stage events that normally occur during the spring. The transducers were removed from wells MS-17, MS-14, and 44 in June 2003 during the environmental-tracer

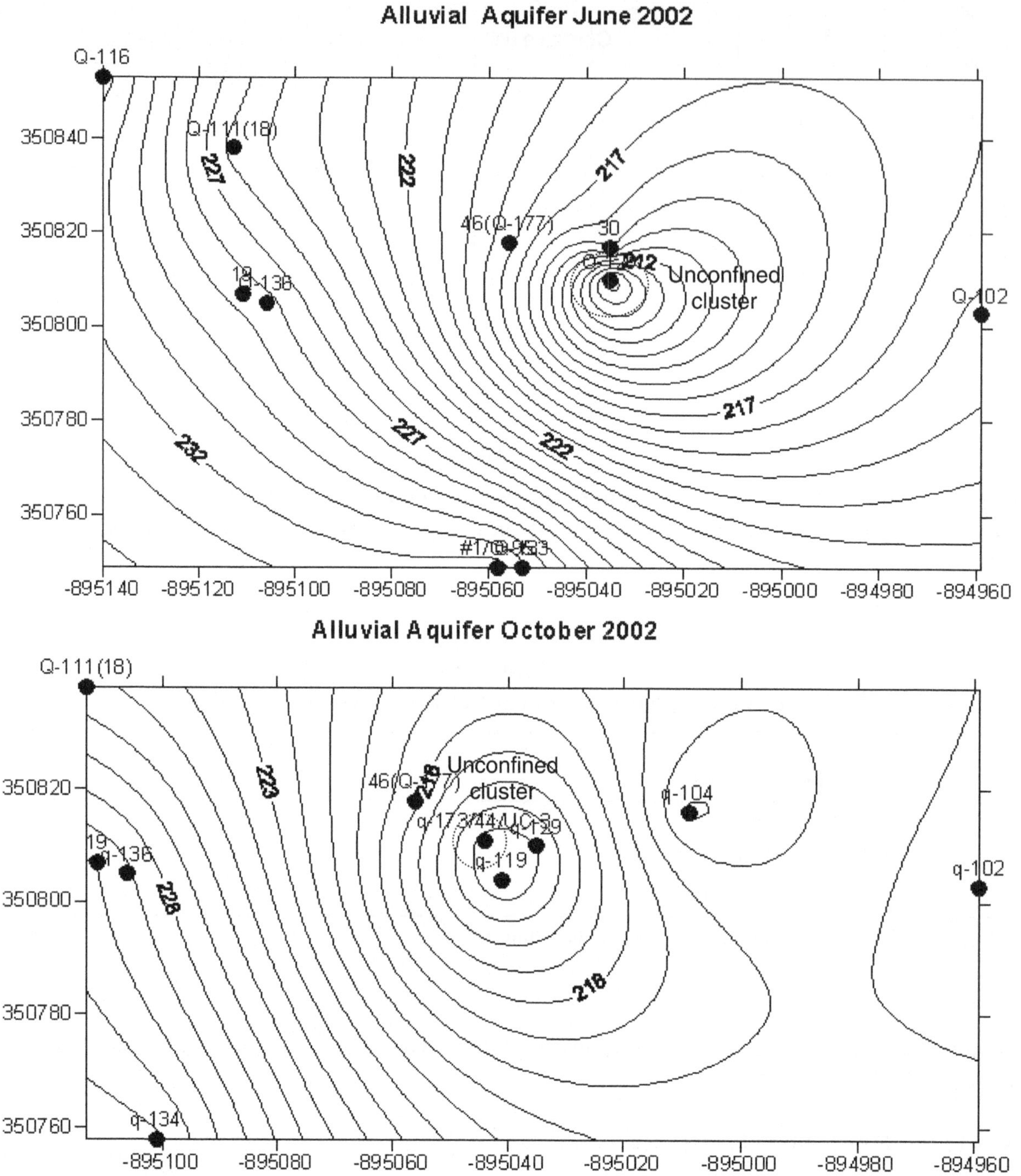

Figure 2.14 Potentiometric contour interpretation for the alluvial aquifer using June and October 2002 water-level data

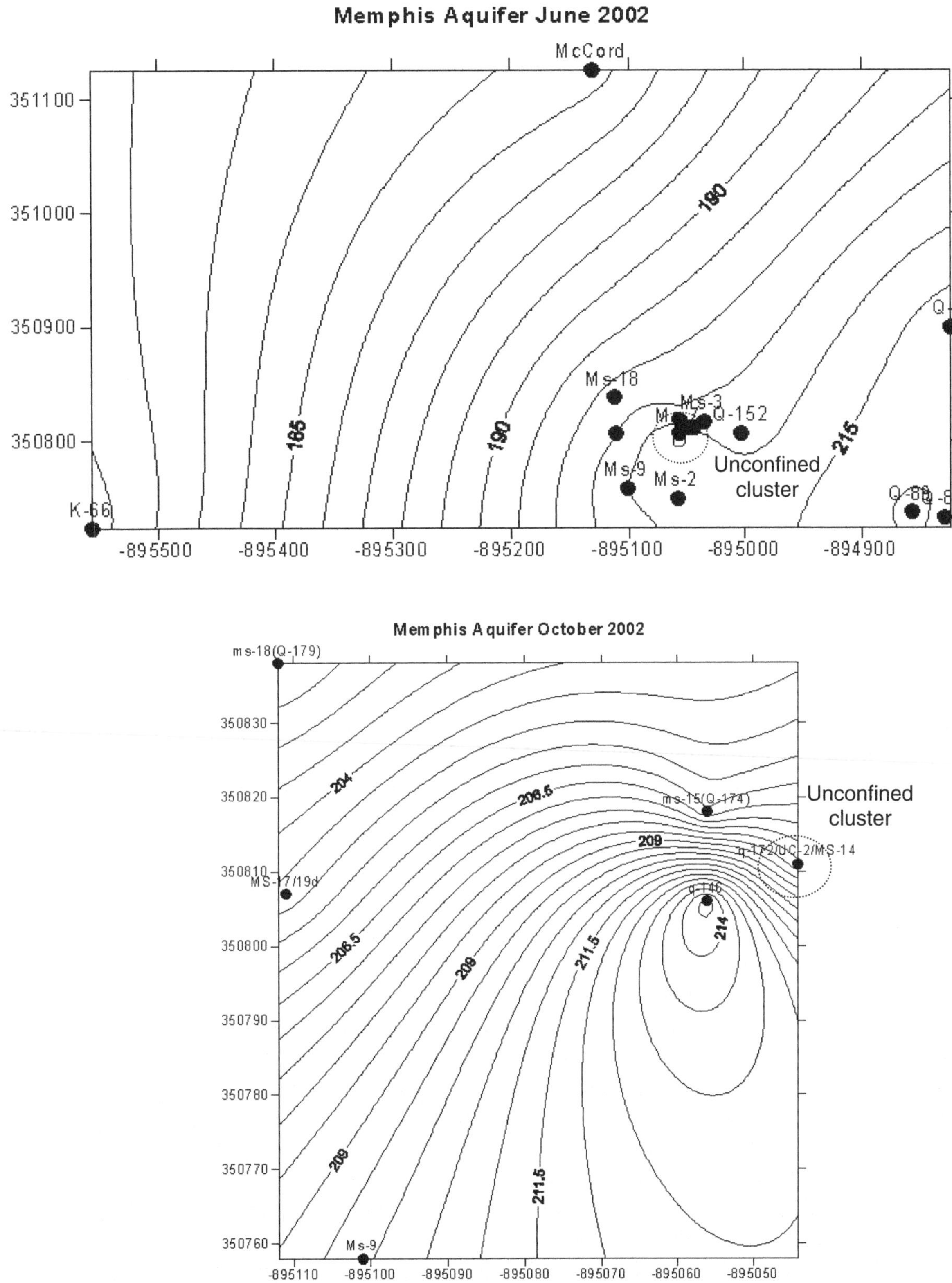

Figure 2.15 Potentiometric contour interpretation for the Memphis aquifer using June and October 2002 water-level data

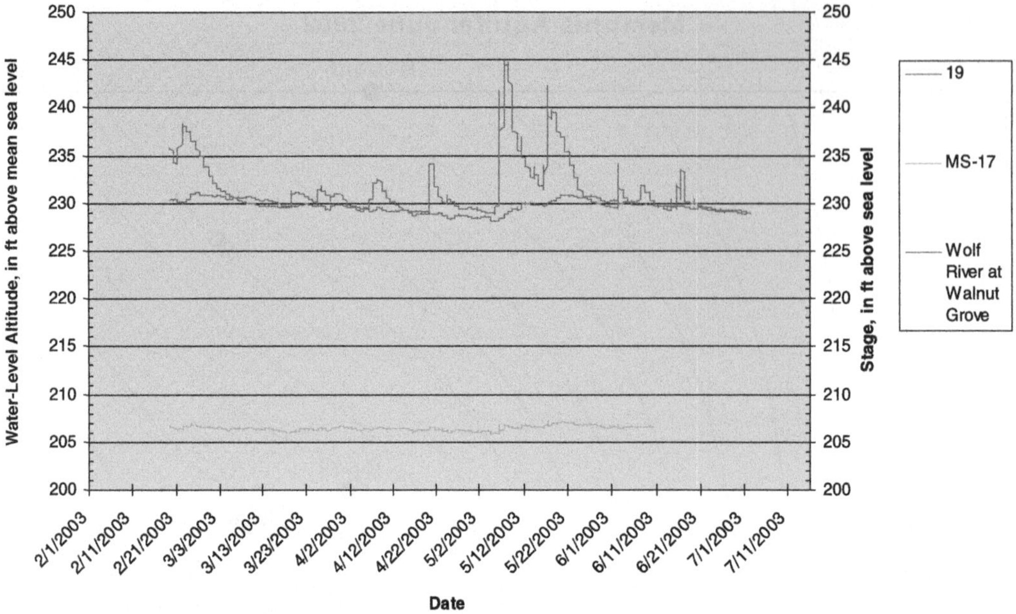

Figure 2.16 Comparison of stage of the Wolf River and water-level altitude in wells 19 and MS-17 (southwestern nest)

sampling event, and from wells 18 and 19 in early July 2003 during a separate visit to the site. A transducer was reinstalled in well 18 in August 2003, along with a transducer to measure and correct the water-level data for changes in barometric pressure (BP), to extend water-level data collected from this well through one annual hydrologic cycle. These transducers were removed during a site visit in May 2004. A transducer also was installed in well 45 at the confined site during the visit in August 2003 to monitor the rate of recovery of water levels in response to bailing this well prior to sampling in December 2003 (see section on Well Hydraulic Testing). This transducer also was removed in May 2004. Water-level data from the wells, and stage data from the Wolf River at the site of a former USGS stream gage at Walnut Grove Road which was estimated from stage data collected at the present USGS Wolf River gage at Germantown Road, are plotted in Figures 2.16, 2.17, and 2.18.

A few key points are indicated from the water-level data. Water levels in wells 19 and MS-17, which are screened in the shallow and Memphis aquifers, respectively, and are located a few hundred feet from the Wolf River (Figure 2.16), vary systematically with Wolf River stage, with the most-pronounced effects seen in well 19 (Figure 2.16). Water levels in the other wells screened in the shallow aquifer (18 and 44) and well MS-14 screened in the Memphis aquifer show little if any correspondence to changes in stage of the Wolf River (Figures 2.17-2.18). Water levels in the shallow wells were lower than the stage of the Wolf River during the period of record (except possibly in well 18 for the period from early July to mid-August 2003), supporting results of previous investigations that indicate that water-level gradients and flow directions in the shallow aquifer are northeast away from the Wolf River and toward the window feature in the study area. Water levels measured in well MS-17 are 25 to 30 feet below those measured in well 19 and the estimated stage of the Wolf River at Walnut Grove Road (Figure 2.16), indicating that the presence of the confining unit at this location influences hydraulic separation and vertical gradients between

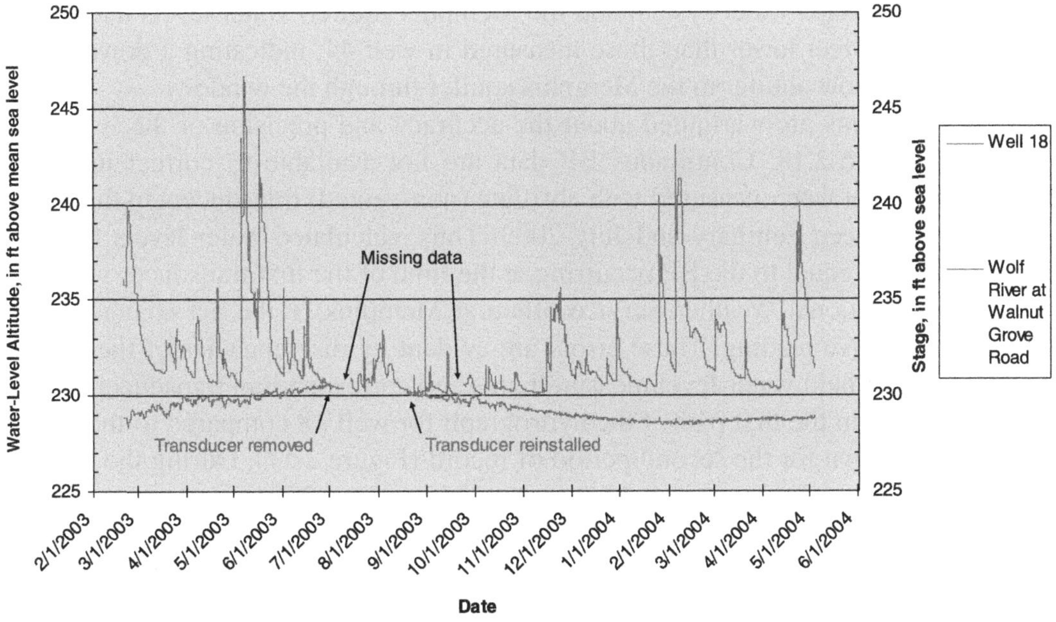

Figure 2.17 Comparison of stage of the Wolf River and water-level altitude in well 18 (northwestern nest)

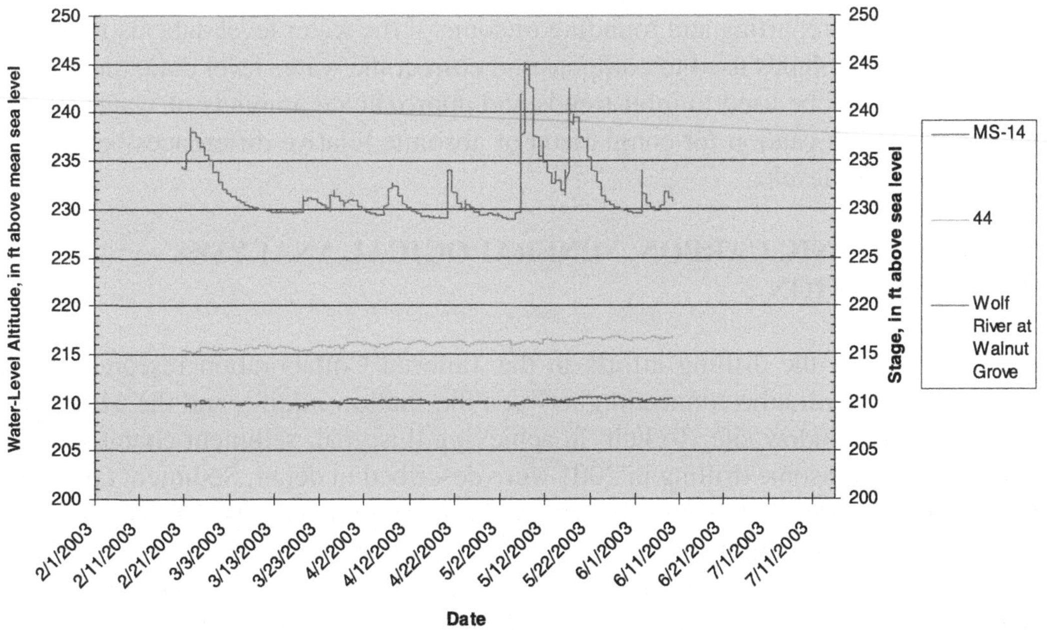

Figure 2.18 Comparison of stage of the Wolf River and water-level altitude in wells MS-14 and 44 (unconfined nest)

the shallow aquifer/surface-water system and the Memphis aquifer. Water levels measured in well MS-14 also are a few feet lower than those measured in well 44, indicating a downward vertical gradient from the shallow aquifer to the Memphis aquifer through the window.

A few comments are warranted about the accuracy and precision of the water-level data shown in Figures 2.16-2.18. Continuous BP data are not available to correct uncompensated groundwater levels that were measured with absolute (non-vented) transducers in the wells for the period of record between February and July 2003. Thus, calculated water levels for this period include errors that are equal to the BP occurring at the time of the first transducer reading (values obtained from the National Weather Service office at Memphis) ± the net change in BP at the times of each successive reading. These errors are evident by discrepancies of the few tenths of feet between periodic field water-level measurements and corresponding transducer readings, and are visible as "noise" in the first part of the hydrograph for well 18 compared to the BP-compensated water levels shown for the second period of record (Figure 2.17). During the second period of water-level data collection from well 18 (August 2003 to May 2004), a transducer deployed in air at the well measured hourly BP values that ranged from 2.09 to 3.46 feet of water (a change of 1.46 feet) and averaged 2.75 feet of water. These BP data were used to correct the second period of water-level record for this well, with agreement between the corrected water levels and periodic measurements to within ± a few hundredths of a foot. Stage of the Wolf River at Walnut Grove Road was estimated for the periods of water-level record by applying a correction of -9.4 feet that was calculated as the difference between the stage of the river during base-flow conditions measured on July 12, 2004, at the present USGS stream gage at Germantown Road and the former stream gage at Walnut Grove Road, located about 3.5 miles downstream. Additionally, differences in the significant figure reporting and rounding of some of the water level data also are a source of error. Because of the methods used to compute and correct the water-level data, the results shown in the hydrographs may be used to infer trends and approximate amounts of water-level change, but should be used with caution for comparison of absolute relative differences between surface-water and groundwater levels.

GRAIN SIZE, ORGANIC CARBON, MINERALOGICAL ANALYSES, AND INTERPRETATION

A chief goal of the drilling efforts in the Tailored Collaboration research project is to better understand the hydraulic connection between the shallow aquifer and the Memphis aquifer at the Shelby Farms window site. To help in achieving this goal, sediment characteristics of the cores obtained from rotasonic drilling in 2001 were described in detail. Sediment column descriptions for rotasonic drilling boreholes were prepared and correlated based on similarity of sedimentary units (lithostratigraphic correlation). The sediment column descriptions were also compared and correlated to columns prepared during previous investigations of the Shelby Farms site (Bradley, 1991; Parks and Mirecki, 1992) to determine the stratigraphic relationships over a broader area.

The rotasonic cores obtained during drilling of boreholes in August and September of 2001 were described in detail in the field. Core sections from borehole MS-13 (0–145 ft), MS-14 (135–145 ft), MS-15 (0–155 ft), and MS-18 (55–155 ft) were packed in pine boxes for storage. The boxes are stored at the University of Memphis in the basement of the Engineering Technology Building. In addition, samples of representative soil and sediments were collected in the field from borehole core MS-17 and the upper 55 feet of borehole core MS-18.

Sampling of the MS-13/-14 and MS-15 cores for laboratory analysis was done at approximately 10-foot intervals. Sampling of the MS-17 and MS-18 cores was done at a somewhat coarser scale and focused on intervals or sediment types not represented in the MS-13/-14 and MS-15 cores. Selected samples were analyzed for their (1) grain-size distribution to better determine the hydraulic conductivity of the sedimentary units, (2) total organic carbon content to better estimate sorptive characteristics, and (3) mineralogy to help determine origin and sorptive characteristics. In addition, selected samples were reacted with 1 Molar KNO_3 solution to assess the cation exchange composition of the sediment. The grain-size characteristics are described using standard stratigraphic nomenclature. The stratigraphy of cores MS-13/-14, -15, -17, and -18 is shown in Figure 2.19. Cores MS-13/-14 are presented as a composite stratigraphic column for the unconfined site because the cores are from borehole separated by approximately 10 feet. The sediments in the cores do not contain shell or body part fossils of animals, but do contain abundant fine plant debris as well as some coal intervals. Four samples were submitted to Dr. Nina Baghai-Riding at Delta State University to assess the pollen and spore content. Initial results indicate abundant angiosperm pollen, some conifer pollen, and less common fern spores. No indication of marine algae (dinoflagellate cysts) is observed in any samples. Given the limitations on evaluating the age of the deposits, correlation to regional stratigraphic units is only approximate and based purely on stratigraphic position, lithologic and hydrologic characteristics. As many as four hydrostratigraphic units are present in the cores: upper alluvium (silt), shallow aquifer, upper Claiborne confining unit, and Memphis aquifer (uppermost part). The upper Claiborne confining unit is absent at the unconfined site; thus, the shallow aquifer and Memphis aquifer are in direct contact. For the purpose of this study, the boundary at the unconfined site is chosen to be at approximately 46 ft.; below this depth the sands are well sorted, lignitic organic chips are common, and pebbles, presumably induced by drilling, are sparsely distributed.

Grain-size data provide information on hydraulic parameters of the sediments as well as information regarding the depositional environments. The grain-size distribution of the fine fraction (<88 micron diameter) was determined using three methods: laser diffraction, pipette method, and hydrometer method. The most reliable results, especially for assessing the grain-size distribution of the fine fraction, are considered to be the hydrometer results. The laser diffraction results appear to have a systematic error in the coarse silt fraction (silt contents appear to be too high). The pipette method provides only 2 data points for the fine fraction (<2 and 62.5 micron contents), but the values are consistent with the hydrometer results. Grain-size statistics based on the sieve (sand fraction) and hydrometer (fine fraction) analysis of samples from each of the borings are given in Table 2.6. Both the pipette and hydrometer results were used to calculate hydraulic conductivity values in Figures 2.12 and 2.13.

The gravel to sand grain size, moderate sorting, and fining-upward grain-size trends in the shallow aquifer are consistent with channel deposits of modern streams in the region. The finer grained upper alluvium is interpreted to be mainly sandy silt overbank deposits from the modern channel, but may also include loess reworked from adjacent hillsides. The upper alluvial deposits have been homogenized by soil processes and contain abundant root tubules and soil glaebules.

The deposits of the upper Claiborne confining unit include two distinct facies: facies A and B. Facies A is interpreted to be deltaic in origin; however, the origin of facies B is less clear. Facies A is predominantly very fine- to fine-grained sand and sandy mud (mixture of clay and silt), generally interbedded at a fine scale. These deposits are dark gray to olive gray, contain copious black organic debris, and have dispersed siderite concretions. The more monotonous, finer grained parts are interpreted to be various delta front and prodelta deposits, most likely

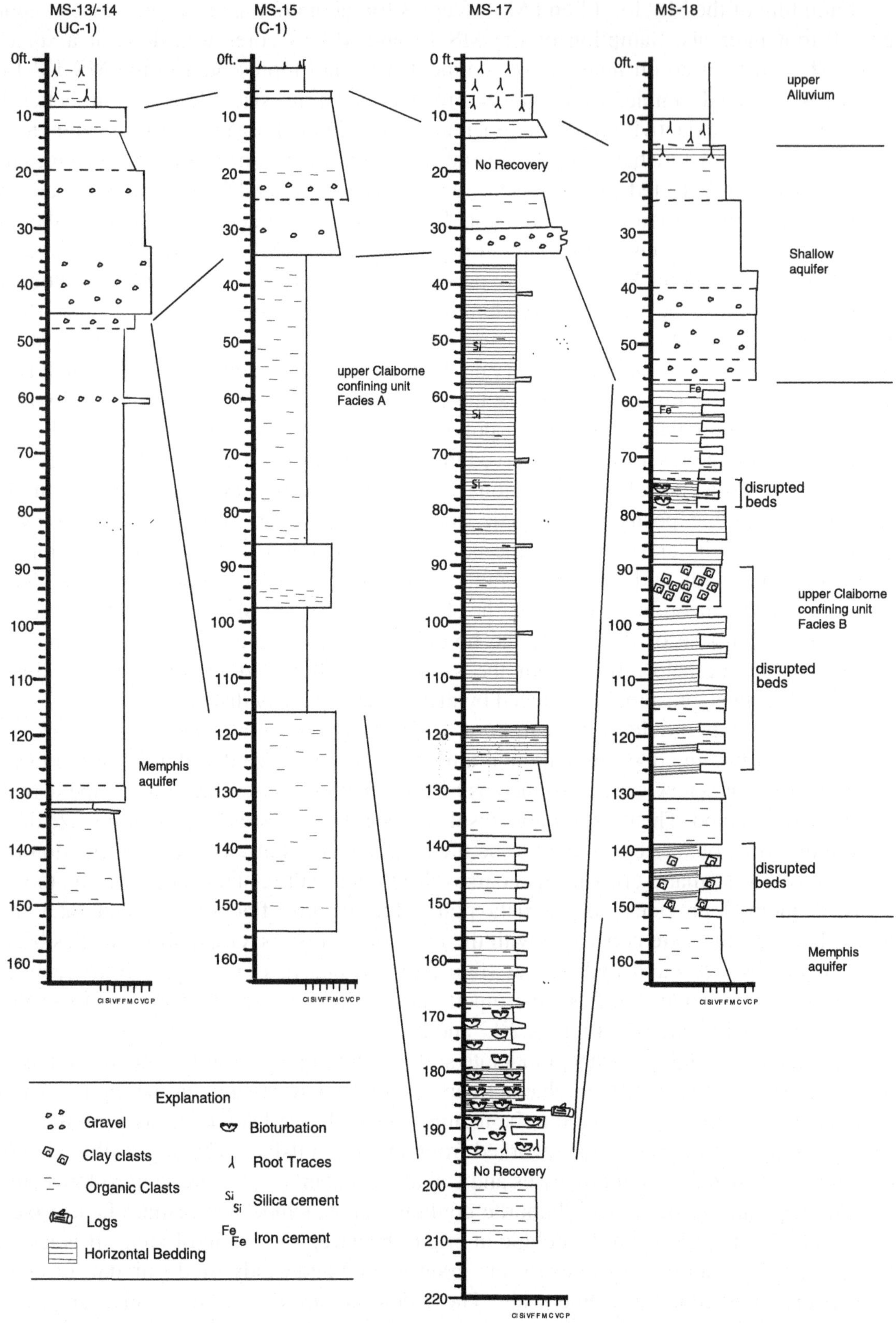

Figure 2.19 Stratigraphic columns from rotasonic drilling at the Shelby Farms site

Table 2.6
Grain size statistics for samples from cores MS-13, MS-15, MS-17, and MS-18. Phi (φ) is the –log2(d), where d is the diameter of grain size in mm.

Core ID	Sample depth	Median φ	Mean φ	Standard deviation σ	Skewness
MS-13	133.0	3.25	3.81	1.55	0.56
MS-13	125.0	1.94	2.12	0.98	3.61
MS-13	86.0	1.91	2.06	0.71	5.76
MS-13	50.0	1.88	2.22	0.92	4.10
MS-13	32.5	1.01	1.80	2.36	0.94
MS-13	7.5	5.23	5.13	2.39	-0.02
MS-13	3.0	4.32	4.98	2.42	0.00
MS-15	125.0	1.94	2.37	1.41	3.65
MS-15	116.0	2.32	3.10	2.10	2.05
MS-15	101.0	4.47	5.26	1.95	0.37
MS-15	86.0	2.59	3.05	1.29	2.75
MS-15	78.0	4.64	5.67	1.81	0.16
MS-15	68.0	4.96	5.68	1.92	0.03
MS-15	55.0	5.87	5.71	1.73	0.05
MS-15	46.0	4.01	5.14	2.00	0.51
MS-15	35.0	5.32	5.21	1.24	-0.71
MS-15	2.5	2.64	3.93	2.17	0.67
MS-17	191.0	3.39	4.08	1.83	0.05
MS-17	167.0	3.32	3.73	1.30	1.03
MS-17	114.0	2.32	2.82	1.50	1.42
MS-17	39.0	3.60	4.24	1.64	-0.09
MS-18	125-135	6.15	5.76	1.66	-0.36
MS-18	143.0	3.32	4.31	1.93	1.09
MS-18	122.0	2.94	3.53	1.43	2.32
MS-18	82.5	3.47	4.69	2.23	0.54
MS-18	67.5	6.15	5.71	2.20	-0.11
MS-18	14.0	5.79	5.41	2.37	-0.07

within a fresh- or brackish-water lake. Similar types of deposits have been observed in core studies from the Mississippi River delta (Coleman and Prior, 1980; Coleman, 1982). Burrows, root traces, and other bioturbation are present in more heterogeneous deposits that include coal and fine- to coarse-grained fining-upward sand intervals. These later deposits are interpreted to be delta-plain marsh, distributary channel, and mouth bar deposits. Again, similar deposits are observed in core from the Mississippi River delta (Coleman and Prior, 1980; Coleman, 1982).

Facies B is predominantly finely interbedded very fine- to fine-grained sand and light gray to white kaolinitic sandy, silty clay, all of which are commonly deformed and brecciated. The brecciation appears different than drilling-induced disruption and is interpreted to be a depositional feature. Facies B contains little organic debris, except in sand-dominated intervals, is oxidized, and also shows little evidence for bioturbation. The most reasonable interpretation is that facies B represents abandoned fluvial channel fill or fluvial-dominated estuary deposits. These deposits are interpreted to be erosionally inset within facies A of the underlying confining unit facies A and Memphis aquifer deposits. The origin of the brecciation is not clear at present, but may represent syndepositional slumping. Confining unit facies B is interpreted to overlie facies A; however, this is based in part on observations elsewhere in the Memphis area. Facies A and B do not occur together in any of our boreholes and interpretations of their distribution in previously drilled boreholes in the study area are based largely on sediment color.

The sands of the upper part of the Memphis aquifer are interpreted to be fluvial or distributary channel deposits. The sands are fine- to medium-grained and well sorted with dispersed sand-sized organic chips. The sands are winnowed of most clay and contain few intraformational clasts. Crossbedding and horizontal bedding are observed rarely; most intervals are massive. The grain-size statistical data were also used to assess the depositional processes responsible for sedimentation of the sand fraction. Using methods presented in Friedman (1982) and Boggs (2001), the standard deviation, skewness, and cumulative frequency data for the sand-rich samples from the Memphis aquifer in cores MS-13 and MS-15 all indicate deposition by stream processes (Figure 2.20). The abundance of organic debris in some sands suggests proximity to a terrestrial source. Although the Memphis aquifer deposits were interpreted as nearshore deposits in a previous report, re-interpretation of the grain-size data, mineralogical data, and regional depositional relationships argue for deposition in slowly aggrading, low gradient stream channels, similar to those in the modern lower Mississippi River valley and Mississippi delta plain.

The detailed stratigraphic columns from the rotasonic cores are lithostratigraphically correlated in Figure 2.21 through Figure 2.24 to stratigraphic columns based on results from earlier AwwaRF test-hole drilling and USGS studies (Bradley, 1991; Parks and Mirecki, 1992). The correlations to previously reported boreholes are made primarily on sediment grain-size and color descriptions. Correlations are presented for borings that penetrated into the confining unit or Memphis aquifer. Cross-section lines CS-2 and -3 show that the contact between the shallow aquifer and the confining unit undulates and gradually deepens toward the Wolf River.

The contact between the Memphis aquifer and the confining unit deepens abruptly in all directions away from the unconfined site. The two relationships demonstrate that the window at the confined site is limited in its extent and is largely a result of heterogeneities in stratigraphic units beneath the shallow aquifer. The bulk mineralogy of all hydrostratigraphic units is dominated by quartz; however, feldspar, micas, and clay minerals are common in the upper alluvium and confining unit samples. In addition, a trace of calcite is present in some upper alluvium samples. Concretions composed of siderite are present in the confining unit facies A, whereas concretions composed of goethite are present in confining unit facies B and the Memphis aquifer. The clay mineral composition and labile organic carbon contents of samples from cores MS-13/MS-14 and MS-15 are presented along with mean grain size and lithological data in Figures 2.25 and 2.26. The labile organic carbon content reflects the most easily oxidized (by hydrogen peroxide) organic carbon in a sample and does not reflect the total organic carbon content. The upper alluvium, shallow aquifer, and Memphis aquifer have relatively low labile organic carbon contents, whereas the highest labile organic carbon contents are present in the

A

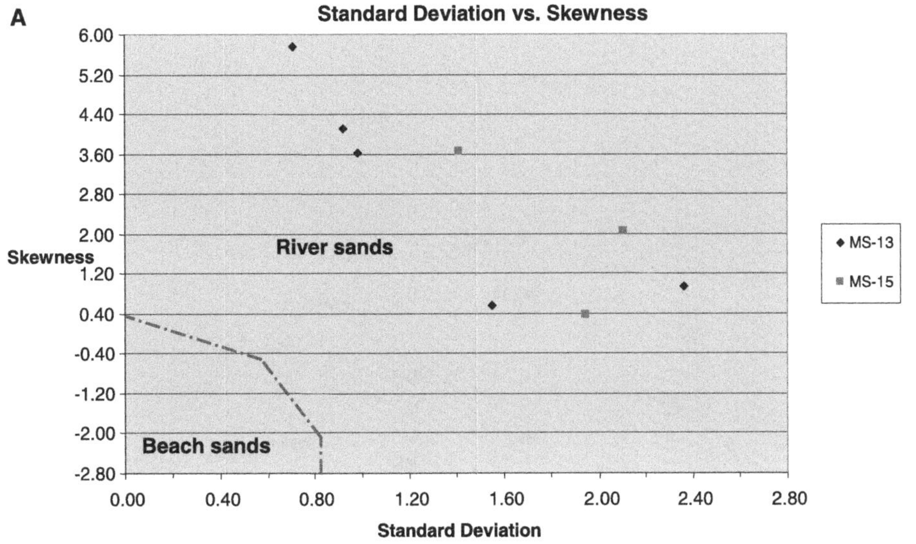

B

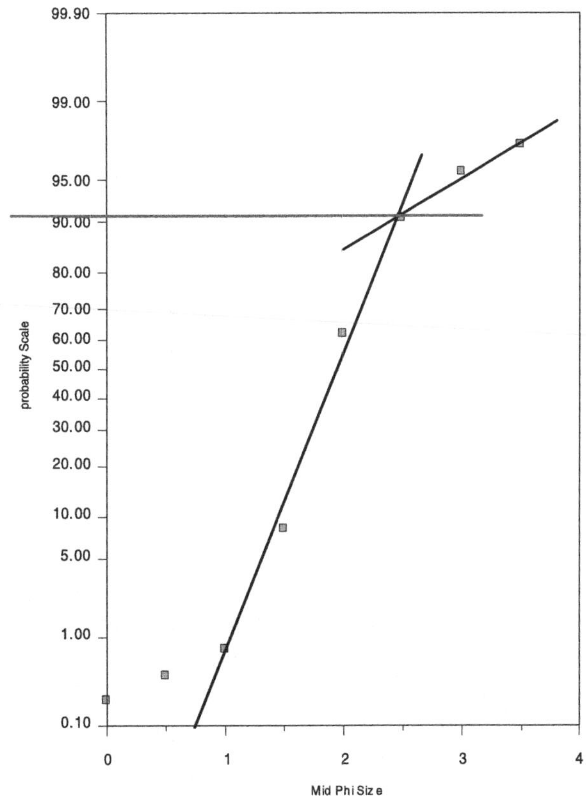

Figure 2.20 (A) Plot of skewness vs standard deviation with fields for beach sands and river sands noted (Boggs, 2001). Data from Memphis aquifer samples in core MS-13 and MS-15. (B) Cumulative frequency probability curve for sample MS-13 -125' illustrating that the suspension grain population comprises approximately 8% of the sample.

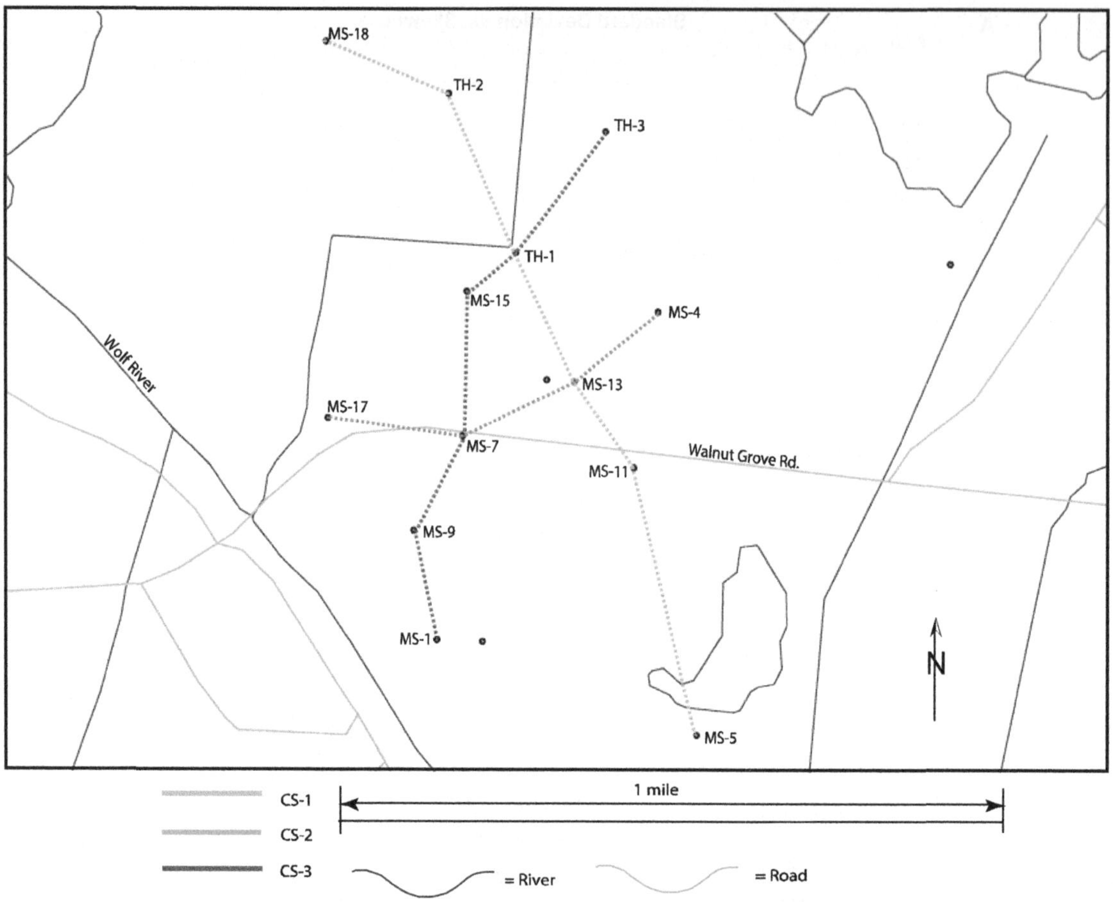

Figure 2.21 Map of Shelby Farms site showing cross-sections

confining unit facies A in core MS-15. The clay mineralogy of the upper alluvium tends to be smectite, kaolinite, and illite dominated, but also contains vermiculite in some samples.

The clay mineralogy of the shallow aquifer is similar, but vermiculite is not present and chlorite is present in some samples. The clay mineralogy of confining unit facies A is variable, but generally dominated by kaolinite and illite with lesser and variable quantities of smectite, chlorite, and kaolinite-smectite mixed-layered clay. The clay mineralogy of confining unit facies B is more homogenous and dominated by kaolinite with lesser quantities of illite and chlorite, and little or no smectite. The clay mineralogy of the Memphis aquifer is also variable and ranges from compositions typical of confining unit facies A to confining unit facies B.

CONCEPTUAL HYDROSTRATIGRAPHIC MODEL

Based on the various types of logging and core data, a conceptual hydrostratigraphic model for the Shelby Farms window site is developed below and shown in Figure 2.27. Five hydrostratigraphic units are defined, each with different hydraulic and stratigraphic properties. The characteristics of each of these units and their potential hydraulic connectivity will influence groundwater flow patterns and geochemical reactions with infiltrating landfill leachate in the study area.

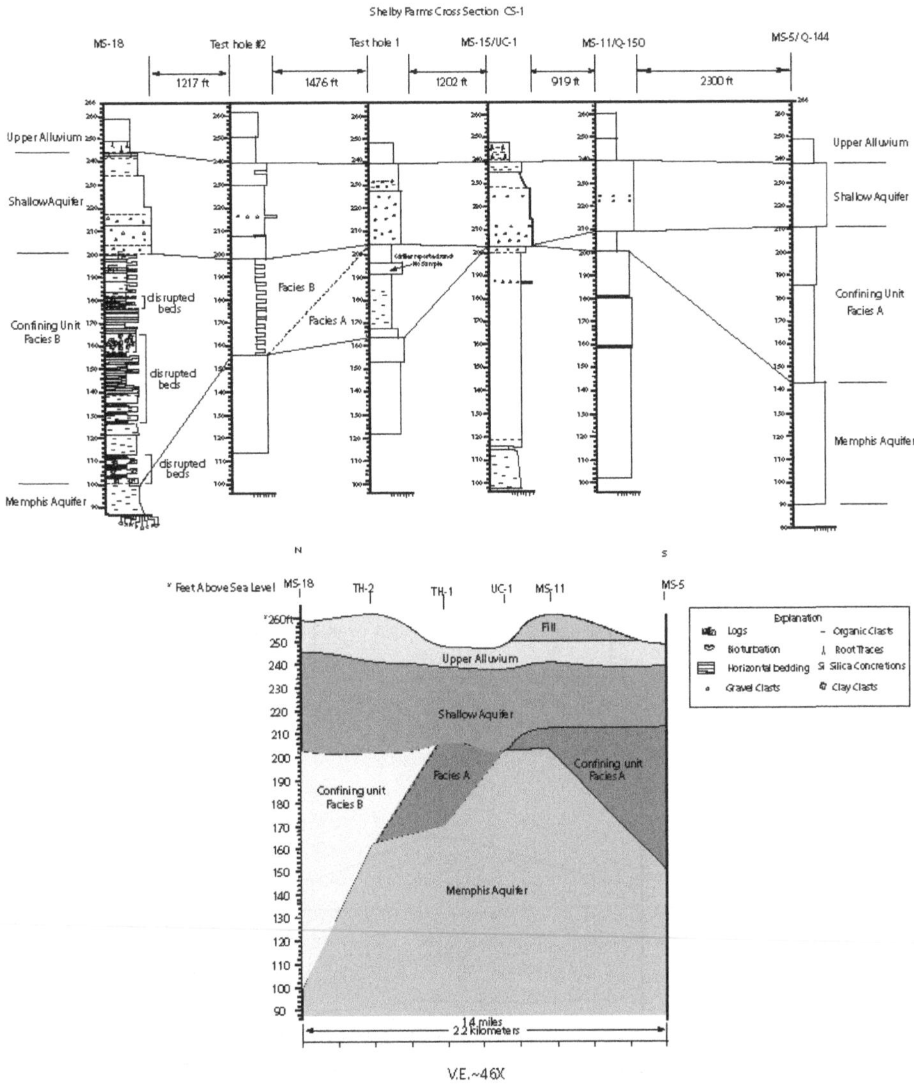

Figure 2.22 Cross-section CS-1 and stratigraphic columns

The upper alluvium is a relatively homogenous unit with low matrix hydraulic conductivity, but numerous root traces and soil-related fractures that may add substantial porosity and permeability. The upper alluvium is also geochemically reactive with abundant clay minerals bearing substantial cation exchange capacity, dispersed iron hydroxide nodules, and dispersed labile organic matter (Larsen et al., 2003). The shallow aquifer has the highest average hydraulic conductivity, although the conductivity generally increases with depth in the unit. The shallow aquifer is somewhat less geochemically reactive than the upper alluvium because of lower overall clay and organic contents.

The confining unit is composed of two differing facies: facies A, a tight, dense dark olive gray silty clay, and facies B, a less dense, white interbedded sand and clay. Although the hydraulic conductivity of facies B was not tested and both facies A and B are below the useful range of grain-size based hydraulic conductivity estimates, facies A is interpreted to have an overall lower and more homogenous hydraulic conductivity than facies B. Interbedded sands within facies B

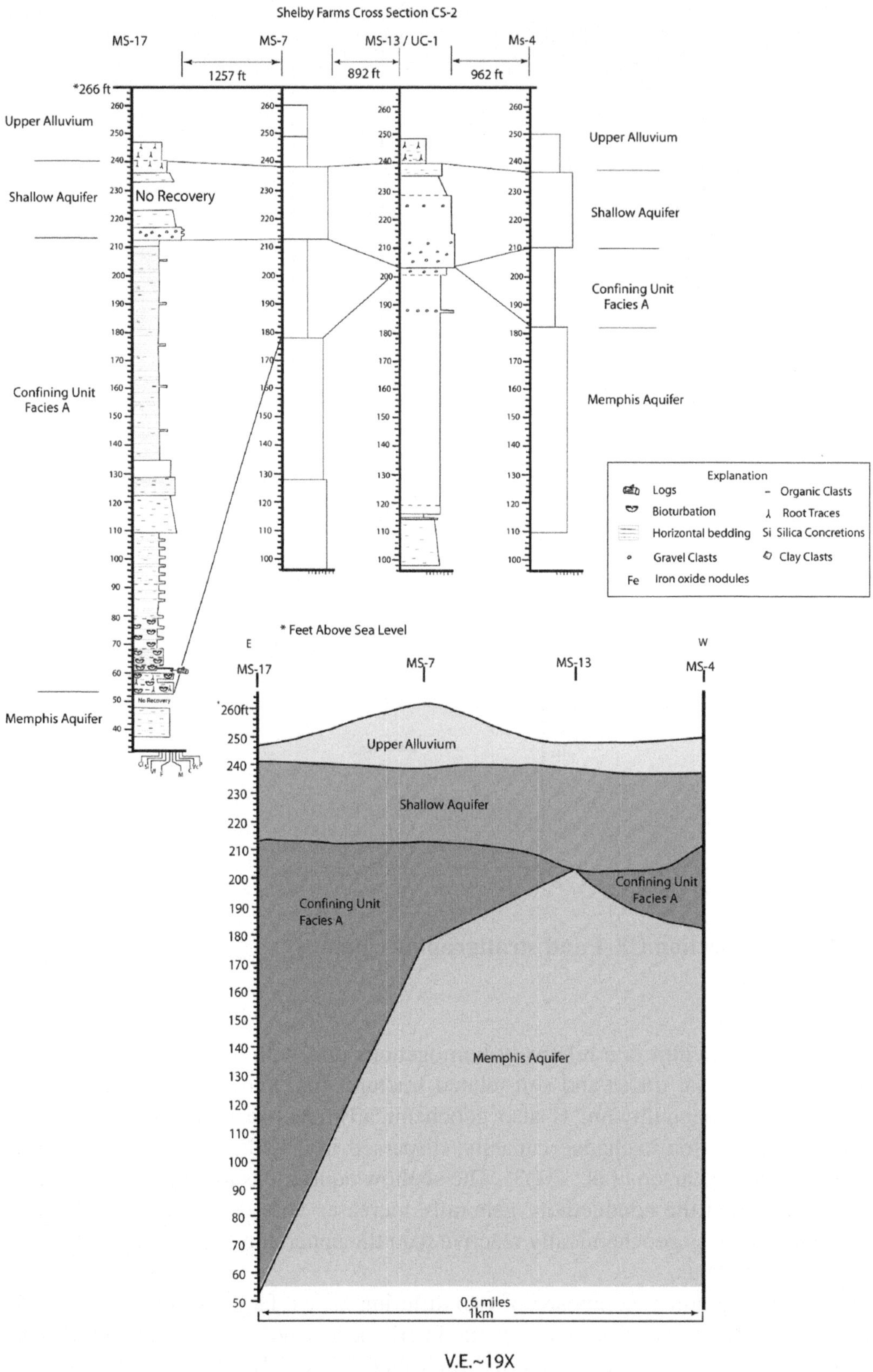

Figure 2.23 Cross-section CS-2 and stratigraphic columns

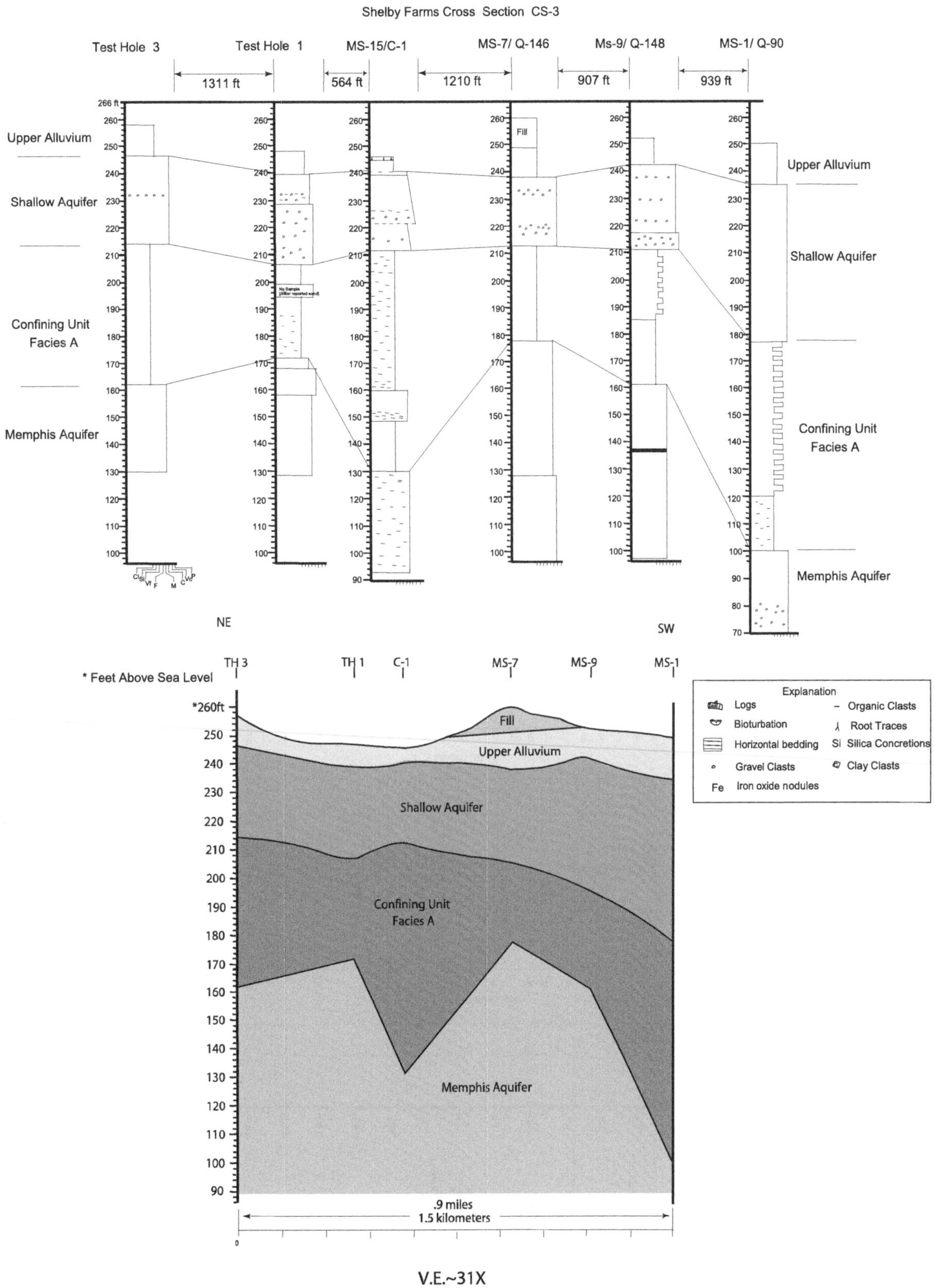

Figure 2.24 Cross-section CS-3 and stratigraphic columns

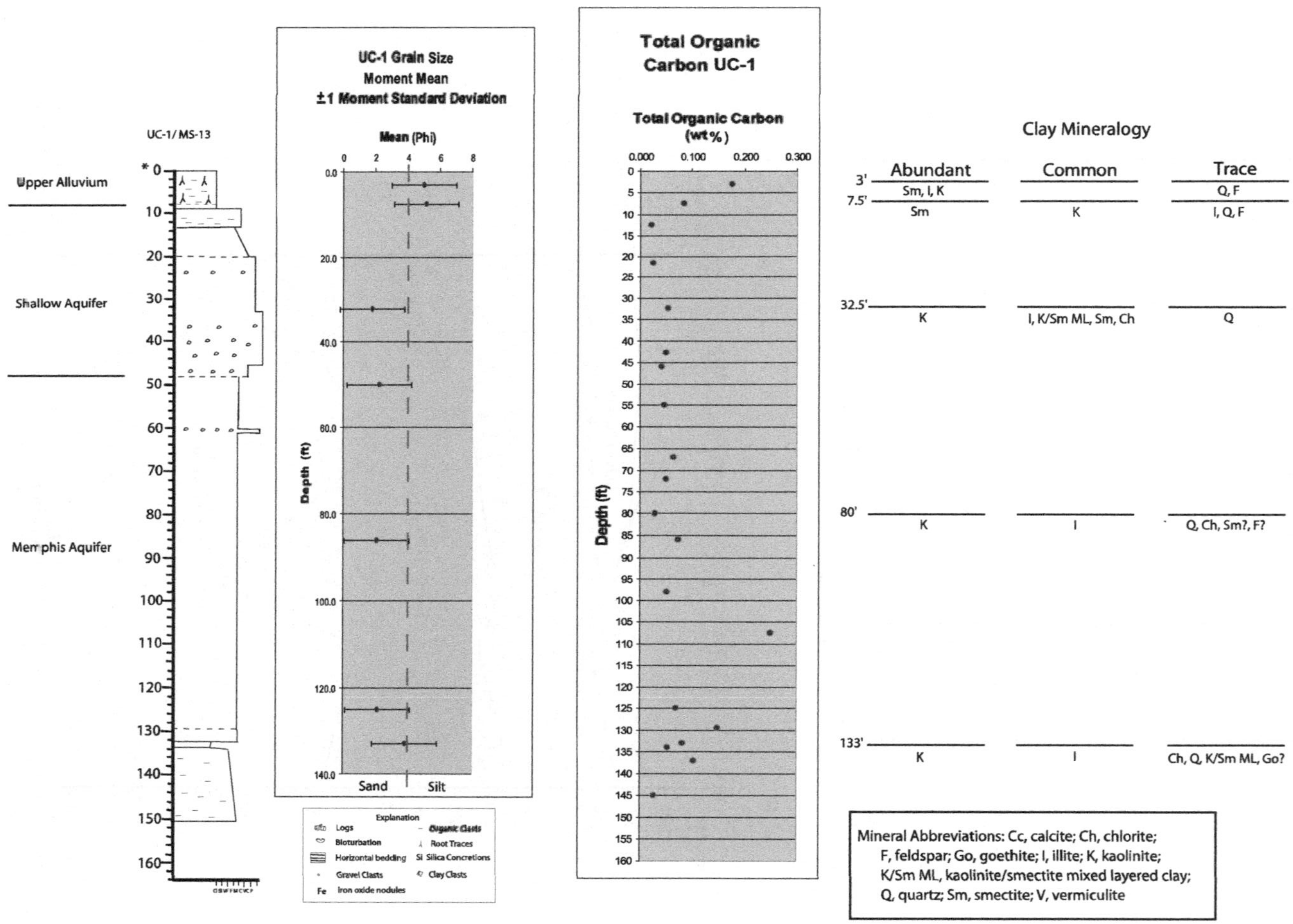

Figure 2.25 Plot of stratigraphic column, mean grain size (sieve-hydrometer data), organic carbon content, and clay mineralogy for core MS-13 (UC-1)

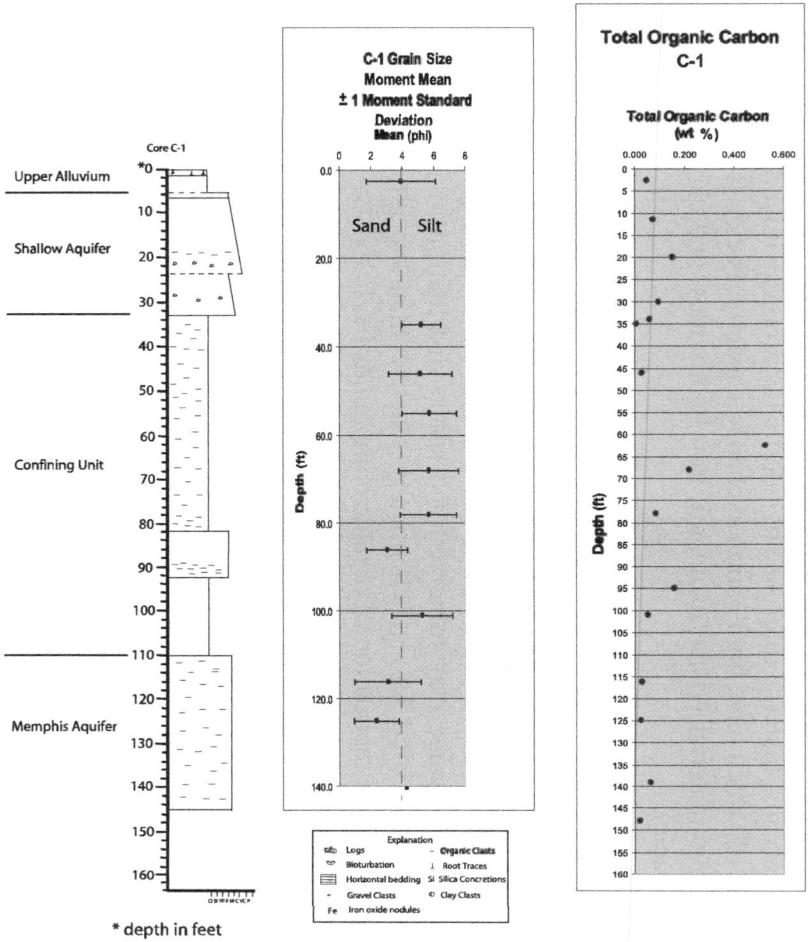

Clay Mineralogy

Depth	Abundant	Common	Trace
2.5'		K	V, I, Q, F, Go, Sm
35'	Sm, K	I	Q
46'	Sm, K	I, Ch	
55'	K, Sm, I		Q, F?, Ch?
68'	K	I	Sm, Q, Ch?
78'	K, Sm	I	Ch, Q, Cc
86'	K	I	Ch, Sm, K/Sm ML
101'	K	I, Ch	Sm, Q
116'	K	I	Q, Ch, Sm
125'	K	Sm	Q, I, Ch, Go?

Mineral Abbreviations: Cc, calcite; Ch, chlorite; F, feldspar; Go, goethite; I, illite; K, kaolinite; K/Sm ML, kaolinite/smectite mixed layered clay; Q, quartz; Sm, smectite; V, vermiculite

Figure 2.26 Plot of stratigraphic column, mean grain size (sieve-hydrometer data), organic carbon content, and clay mineralogy for core MS-15 (C-1)

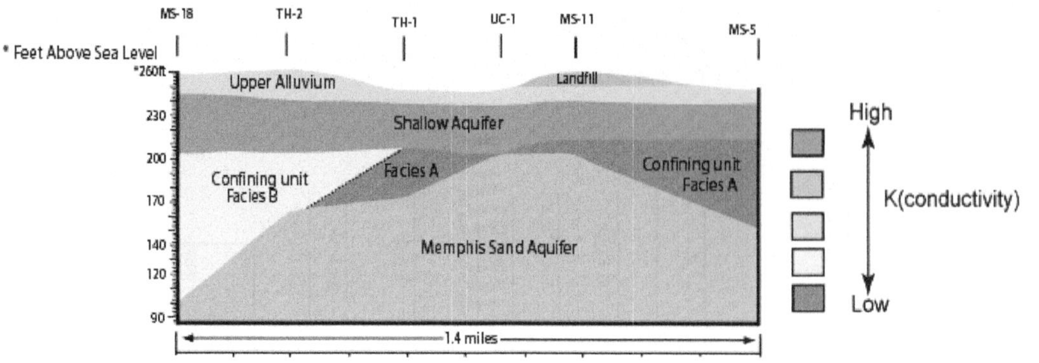

Figure 2.27 Hydrostratigraphic conceptual model of the Shelby Farms study area

contain substantial quantities of clay, but the overall electrical conductance values are much greater than facies A (resembling higher conductivity sand units) and the facies B deposits are overall less compacted than facies A. The distribution of these two facies is only partially understood in the study area and, based on observations in other areas within the region, complex stratigraphic relationships are likely to exist. Furthermore, the distribution and continuity of sand intervals within the confining unit are not well understood based on the available stratigraphic information. The geochemical reactivity of facies A is much greater than that of facies B because of the abundance of organic material, presence of siderite (a carbonate mineral), and abundance of clay minerals, some of which have high cation exchange capacity. Facies B contains a relatively low content of organic material (visual estimate), iron hydroxides, and clay minerals with high cation exchange capacity.

The Memphis aquifer has a relatively high hydraulic conductivity, although somewhat lower in value than that of the shallow aquifer. The slug test data suggest a more heterogeneous distribution of hydraulic conductivity values than the grain-size data, but overall the distribution of values is more homogenous than those of the confining unit or shallow aquifer at the study site. The Memphis aquifer does contain dispersed organic material, but has a low clay content (all of which are low cation exchange clay minerals) and spotty distribution of iron hydroxides. The Memphis aquifer is considered less geochemically reactive than the other hydrostratigraphic units, although potential for retardation of organic compounds likely exists.

CHAPTER 3
ENVIRONMENTAL TRACERS

SAMPLE COLLECTION METHODOLOGY

Prior to collection of samples for environmental tracers and geochemistry, the wells were purged for a total of three to five well volumes and until general chemical parameters (pH and conductivity) were stable to within ±5% for consecutive readings. The purging process was performed with a 1.8-inch Grundfos pump set at a low enough pumping rate to not induce large head changes at the wellhead. Purging began at the top of the water column, and the pump was slowly lowered until it was at the top of the well screen. This procedure was used to prevent the close-fitting pump in the 2-inch wells from acting as a piston and pushing well-bore water through the well screen into the formation.

To facilitate the purging procedure, and meet the special requirements for krypton-85 (^{85}Kr) sampling, significant changes had to be made to the available Grundfos pump system. As the predominant source of ^{85}Kr in groundwater is from dissolution of atmospheric gases, it is very important to prevent any contact between water collected for ^{85}Kr analysis and the atmosphere. This is especially true when considering that the 100-fold increase in atmospheric concentration of ^{85}Kr over the past 40 years, combined with its 10.8 year half-life, results in the isotopic abundance (ratio of ^{85}Kr to stable krypton isotopes) in modern air being more than a factor of 1,000 greater than in a 40-year old groundwater sample.

The sampling tube supplied with the Grundfos pump, which is permeable to air, was replaced with 200 feet of ½" low gas-permeability Nylon-6 tubing. As this tubing is extremely stiff, it had to be wound (together with the pump power cable) onto a custom 24" diameter drum. To permit operating the pump for purging while lowering it into the well, the drum was fitted with a hermetic rotary water joint and a custom electrical commutator to power the pump. A manifold was built to control and monitor pump discharge pressure, eliminate any dead-space where air or residual water could be trapped, facilitate sampling for tritium/helium (^{3}H/^{3}He), geochemistry and ^{85}Kr, and route purged water away from the well-head area. The sampling set-up is shown in Figure 3.1.

After purging, samples were collected for ^{3}H/^{3}He, geochemical analysis, and ^{85}Kr. Samples for geochemical analysis included a 250-mL raw water sample for anion and titration analysis, and 250-mL filtered (.45 micron filter), acidified (1 mL concentrated nitric acid) sample for cation and metal analysis. Duplicates and field blank samples were taken during at least one sampling event. ^{3}H/^{3}He samples were collected by filling a 1-liter amber Boston round glass bottle or a 1-liter polyethylene bottle.

^{85}Kr sampling was performed subsequent to the collection of the ^{3}H/^{3}He samples. The sampling consisted of filling 20-liter evacuated glass bottles while keeping the well-head discharge pressure above 15 pounds per square inch (psig) to avoid degassing of the water in the sampling line before it entered the sampling bottle. Subsequent to the collection of the ^{85}Kr samples, diffusion samplers developed by Dr. Kip Solomon were lowered into the well screen and allowed to equilibrate for over one week to collect a representative concentration of dissolved gases (He, N_2, Ne, Ar, and Kr) at equilibrium conditions. Following retrieval of the diffusion samplers, a total dissolved gas pressure measurement was made in the well at the depth that the sampler was suspended.

Figure 3.1 The assembled bailer before deployment into well 45 by Jack Carmichael. The thin-walled 1-¾" O.D. stainless steel tube holds 2.2 liters of water between upper and lower valves. Note details on right: Double bailer attachment (top). The single chain going through center is connected to both upper and lower valves; the yellow rope attaches to the bailer chamber with three short chains. Bailer is lowered into well (center) by center chain, which keeps both valves open, and water in well flows through chamber. Bailer is retrieved by yellow rope (keeping center chain slack), which closes both valves. Lower valve and 12" rods which keep bailer out of sump (bottom).

Three sampling events were conducted: November 2002, June 2003, and December 2003. During the November 2002 sampling event, samples for $^3H/^3He$, geochemistry, and ^{85}Kr analysis were collected from wells MS-13, MS-14, 44, and MS-12 from the unconfined cluster, well MS-16 from the confined cluster, and wells MS-17 and MS-18, also in the confined region but on either side of the confined cluster. During the June 2003 sampling event, the same set of wells were sampled as in November 2002. Samples were collected for $^3H/^3He$ and ^{85}Kr analysis, both as a reproducibility check, and to look for seasonal variability. A final round of sampling was conducted in December, 2003 in which wells MS-15 (C-1), 45 (C-3), 46 (C-4), and an up-gradient Memphis aquifer well (Patriot Lake well) were sampled for $^3H/^3He$ and geochemistry. Well MS-15 and the Patriot Lake well were also sampled for ^{85}Kr. The Patriot Lake well is used by Shelby Farms personnel to maintain water levels in Patriot Lake, a recreational lake in the park, and is utilized only on rare and intermittent occasions. Sampling procedures for the wells were generally similar to those used in November 2002, except for sampling of well 45, which is screened in the confining unit. Because of slow recovery, the ^3He diffusion cells were emplaced in well 45 in October, after the well was partially bailed. The diffusion cells were attached to a specially designed bailer, designed by Dr. Norbert Thonnard (Figure 3.1). The bailer was extracted in December after having resided in the recovering well for over a month; thus, allowing for the collection of a representative sample for geochemical and dissolved gas analyses at the University of Utah and geochemical analyses at the University of Memphis.

$^3H/^3HE$ RESULTS

The raw water samples in the amber glass bottles were packed in a cooler and shipped to the University of Utah Dissolved Gas Laboratory for ^3H analysis by the helium in-growth method. In this method, ^3H activity was determined by the amount of ^3He accumulated in a sample over a series of 3 to 4 months (Clarke et al., 1976; Bayer et al., 1989). The practical detection limit for the helium in-growth method is 0.05 tritium units (TU) (Solomon and Cook, 2000). The percent error ranges from approximately ±10% at 1 TU to ±70% at the detection limit.

Following retrieval, cold-sealing, and packing in Nalgene bottles, the gas diffusion samplers were sent to the Dissolved Gas Laboratory at the University of Utah. The gas isotope contents of the samplers are determined using a line based upon the design described in "A Static Mass Spectrometer With Triple Collection for Nitrogen and Neon Isotopes," (Craig 1993). The system is designed to measure and remove atmospheric gas components using a combination of SAES getter technology and cryogenics. The system is held at high vacuum using a combination of rough pumps, turbo getters and ion pumps. The entire system is controlled through two computer systems utilizing National Instruments hardware and software.

Isotopic measurements are made using the following:

- Stanford Research SRS–Model RGA 300 quadrapole mass spectrometer. Measurements for all gases except helium are made using this instrument. This includes: $^{28}N_2$, ^{36}Ar, ^{40}Ar, ^{20}Ne, ^{22}Ne, ^{78}Kr, ^{84}Kr, ^{86}Kr, ^{126}Xe, ^{129}Xe, ^{131}Xe, ^{132}Xe, ^{134}Xe, ^{136}Xe.
- Mass Analyzers Products – Model 215-50 Magnetic Sector Mass Spectrometer. This is used to measure ^3He and ^4He.

Table 3.1
Gas analysis detection limits and precision, as well as other characteristics of measurement process

Constituent	AWRL* (cc/g stp)†	RL (lab reporting limit)	MDL (method detection limit) (cc/g stp)	Extraction efficiently (%)	Precision (%)
Nitrogen	1.5×10^{-6}	1% of AWRL	1.5×10^{-8}	99.5	5.0
Argon	5.4×10^{-8}	1% of AWRL	3.0×10^{-10}	99.5	3.0
Neon	7.9×10^{-7}	1% of AWRL	1.0×10^{-9}	99.5	2.0
Krypton	3.4×10^{-8}	1% of AWRL	5.0×10^{-10}	99.5	4.0
Xenon	2.2×10^{-9}	1% of AWRL	2.4×10^{-10}	99.5	4.0
Helium	4.2×10^{-9}	1% of AWRL	1.0×10^{-10}	99.5	1.0

* Ambient water reporting limit (typical air concentrations).
† cc/g stp – cubic centimeters at standard temperature and pressure.

Standardization is completed daily using a tank of "dry" atmosphere as a standard. All analytical calibrations are made using this standard. In addition, air equilibrated water samples collected at different temperatures are analyzed as blind standards. Typically, four standards are analyzed for six unknown samples in a given day. The daily schedule is: two standards are run in the morning to ensure proper communication and re-center the peaks, then three unknown samples are run followed by a standard. If the standard is acceptable, three more samples are analyzed which are followed by a closing standard. In addition, one air equilibrated water sample is analyzed for approximately every ten unknown samples. The method detection limits and typical analytical precision are given in Table 3.1.

During an analysis, all pertinent information is collected by computer and imported to a Microsoft Excel worksheet. Three files are produced for every one analysis. On a daily basis, these files are "compiled" into a single workbook using Visual Basic code developed in the Noble Gas lab. This code incorporates the daily standard runs and produces a final sample report. These reports are then reviewed by hand and used in the age-calculation process.

Groundwater age calculations using the $^3H/^3He$ method typically require (1) correcting for excess air (concentrations in excess of equilibrium at air-saturation) dissolved in the sample, (2) correcting for the presence of ^3He due to processes other than ^3H decay, and (3) calculating the tritiogenic ^3He content. Excess air is thought to be a result of partial dissolution of air bubbles caught in porous media during rising water-table conditions and subsequent incomplete diffusive equilibrium (Stute and Schlosser, 2000). As a result, dissolved noble gas contents in groundwater are a complex function of temperature, recharge rate, and degree of atmospheric equilibration. Furthermore, atmospheric noble gases fractionate from one another during partial gas bubble dissolution and subsequent diffusive equilibration. To account for excess air, a least-squares gas dissolution model is applied to as many noble gas concentrations as are available and the errors are minimized using a chi-squared function. The model applied to the data is a modified form of the closed-equilibrium model developed by Aeschbach-Hertig et al. (2000). The resulting recharge temperature provides a best estimate for equilibration conditions with the atmosphere.

^3He dissolved in groundwater arises from four sources: atmospheric equilibration, ^3H decay, shallow subsurface nuclear reactions, and the mantle (Solomon and Cook, 2000). The atmospheric component can be corrected for by using an excess air model to determine recharge temperature and then applying the appropriate gas-water equilibrium constants. Shallow subsurface nuclear reactions generate ^3He from fission of ^6Li produced during U- and Th-series decay. The resulting $^3He/^4He$ ratio from such reactions is approximately 1×10^{-8}, which is two orders of magnitude less than that of the atmosphere (1.36×10^{-6} at 10°C). Relatively large amounts of ^4He from shallow subsurface reactions must be present for this to be an important factor. Nonetheless, ^3He from shallow subsurface reactions can be estimated using $^3He/^4He$ ratio of terragenic (crustal–produced) helium and using neon to correct for atmospheric ^4He contributions. Mantle helium has a $^3He/^4He$ ratio approximately 10 times greater than that of atmospheric helium; thus, small contributions of mantle helium may be difficult to detect. Given the shallow groundwater flow paths inferred by groundwater flow modeling in the Memphis area (Brahana and Broshears, 2001) and the cratonic tectonic setting, mantle-sourced helium is not likely to contribute to the helium present in shallow Memphis aquifer waters.

The $^3H/^3He$ age of a groundwater sample is defined as:

$$t = \lambda^{-1} \times \ln(^3He_t/^3H+1) \qquad (3.1)$$

where λ is the tritium decay constant, ^3He_t is tritiogenic ^3He, and t is the age in years of the modern (<50 years old) component of the water. The noble gas isotope contents, R/R_a, ^3He_t, ^3H, recharge temperature, and other aspects of the age calculations for the Shelby Farms samples are given in Table 3.2. The ratio R/R_a measures how closely the measured helium-isotope ratio is to that of air-saturated water. The R/R_a values approach 1 for young samples (i.e., well 44) and have progressively higher ratios for older samples, which is consistent with domination of atmospheric helium contributions. Terragenic ^4He is one to two orders of magnitude less than the total ^4He, suggesting a minor crustal ^3He component is present. Excess air is present in some samples but not others and shows no consistent relationship to groundwater age or depth of sample. Several samples did not fit the excess-air gas model, resulting in high residual errors. Given the potential losses of gas during sampling, possible gas stripping at depth, and other sources of qualitative error, the total error on the age estimate of a given sample is difficult to assess. Considering that all possible measures were taken to minimize error and gas loss, it is reasonable to assume based on past sampling experience and replicate analyses that the sum of the errors is less than 10% of the age determination.

Another assessment of the quality and interpretation of the age data is to compare the age results against expected ^3H loadings for a given recharge year. ^3H concentrations in the atmosphere have not been measured in the Memphis area and the nearest station is at St. Louis, Missouri. ^3H concentrations vary in relation to weather patterns, seasonally and by latitude. An extrapolated history of ^3H activity in the atmosphere was prepared by Ivey (2003) using the code developed by Cross (1989). The extrapolated ^3H activity history compares favorably with that of Cape Hatteras, North Carolina, which is located at approximately the same latitude as Memphis, suggesting that the errors are probably minimal. Assuming total retention of dissolved helium in water that recharged in a given year, the sum of the tritiogenic helium and ^3H activities should approximately equal the ^3H loading for that year. The ^3H activity in the atmosphere varies seasonally, so the yearly average, based on monthly values, and one standard deviation of the mean were determined. The results are presented in Table 3.3. Given the variations in recharge and ^3H

Table 3.2
Environmental tracer data from 2002 and 2003 sampling events

November 2002 Dissolved Gases

Sample ID	Run ID	N_2 (ccSTP/g)	^{40}Ar (ccSTP/g)	^{84}Kr (ccSTP/g)	^{20}Ne (ccSTP/g)	^{4}He (ccSTP/g)	R/Ra*
MS-13	03060305	0.0125	2.53E-04	2.89E-08	1.51E-07	4.07E-08	1.143
MS-14	04080306	0.0120	1.87E-04	1.98E-08	1.13E-07	3.12E-08	1.198
MS-12	04080305	0.0148	3.03E-04	3.95E-08	1.70E-07	4.67E-08	1.059
44	04080304	0.0166	3.91E-04	5.13E-08	2.24E-07	5.92E-08	0.994
MS-16	03070306	0.0157	4.03E-04	4.64E-08	8.70E-08	4.02E-08	1.728
MS-17	03060303	0.0150	2.69E-04	3.25E-08	1.31E-07	3.42E-08	1.592
MS-18	03070303	0.0155	3.54E-04	4.85E-08	1.61E-07	4.82E-08	1.560

November 2002 Tritium/^{3}Helium and Associated Apparent Ages

Sample ID	Well depth (ft -BLS)	$^{3}He_t$ using Ne (TU)	^{3}H (TU)	Apparent age using Ne (yr)	Terragenic ^{4}He using Ne (ccSTP/g)	Excess air (ccSTP/g)
MS-13	116	3.2	5.1	8.9	nd†	0.00
MS-14	149	4.9	2.4	19.8	2.7E-09	0.00
MS-12	87.5	2.5	4.5	8.0	1.9E-09	6.73E-04
44	50	nd	3.6	na‡	nd	1.21E-3
MS-16	97	28.2	5.4	32.9	2.3E-08	0.00
MS-17	166	11.2	3.4	26.2	nd	0.00
MS-18	142	19.2	3.9	31.8	8.1E-09	0.00

June 2003 Dissolved Gases

Sample ID	Run ID	N_2 (ccSTP/g)	^{40}Ar (ccSTP/g)	^{84}Kr (ccSTP/g)	^{20}Ne (ccSTP/g)	^{4}He (ccSTP/g)	R/Ra*
MS-13	01060408	0.0120	2.53E-04	2.89E-08	1.57E-07	4.66E-08	1.154
MS-14	01060409	0.0106	2.29E-04	2.72E-08	1.45E-07	4.18E-08	1.194
MS-12§	na	na	na	na	na	na	na
44	12170304	0.0169	3.73E-04	4.56E-08	2.17E-07	5.81E-08	1.016
MS-16	01060403	0.0154	3.48E-04	4.44E-08	1.88E-07	5.23E-08	1.530
MS-17	12170303	0.0127	3.02E-04	4.02E-08	1.55E-07	4.03E-08	1.495
MS-18	11100303	0.0149	3.71E-04	4.52E-08	2.00E-07	5.21E-08	1.410

(continued)

Table 3.2 (Continued)

June 2003 Tritium/^{3}Helium and Associated Apparent Ages

Sample ID	Well depth (ft -BLS)	^3He_t using Ne (TU)	^3H (TU)	Apparent age using Ne (yr)	Terragenic ^4He using Ne (ccSTP/g)	Excess air (ccSTP/g)
MS-13	116	6.1	5.7	13.1	4.1E-09	0.00E+00
MS-14	149	6.1	6.7	11.7	3.2E-09	0.00E+00
MS-12§	87.5	na	na	na	na	na
44	50	0.4	3.6	1.7	nd	2.57E-03
MS-16	97	16.6	5.3	25.4	2.4E-09	1.01E-03
MS-17	166	10.8	4.9	21.0	nd	0.00E+00
MS-18	142	11.4	4.3	23.2	nd	1.48E-03

December 2003 Dissolved Gases

Sample ID	Run ID	N_2 (ccSTP/g)	^{40}Ar (ccSTP/g)	^{84}Kr (ccSTP/g)	^{20}Ne (ccSTP/g)	^4He (ccSTP/g)	R/Ra*
MS-15 (C-1)	02020409	0.0134	3.22E-04	4.07E-08	1.70E-07	4.62E-08	1.569
45 (C-3)	02020405	0.0230	5.37E-04	6.08E-08	2.93E-07	7.75E-08	0.942
46 (C-4)	02020408	0.0142	3.67E-04	4.99E-08	1.99E-07	5.27E-08	1.069
Patriot Lake	02020411	0.0146	3.65E-04	4.78E-08	1.92E-07	5.21E-08	1.021

December 2003 Tritium/^{3}Helium and Associated Apparent Ages

Sample ID	Well depth (ft-BLS)	^3He_t using Ne (TU)	^3H (TU)	Apparent age using Ne (yr)	Terragenic ^4He using Ne (ccSTP/g)	Excess air (ccSTP/g)
MS-15 (C-1)	157	15.5	5.6	23.8	1.7E-09	0.00E+00
45 (C-3)	61	-2.9	0.1	>40 years	7.2E-10	6.08E-03
46 (C-4)	36	2.3	3.5	9.1	4.9E-10	1.15E-03
Patriot Lake	350	1.6	0.6	23.2	2.0E-09	8.50E-04

* R is the $^3He/^4He$ ratio of the sample. Ra is the $^3He/^4He$ of the air standard (1.384E-6).
† nd = non-detectable.
‡ na = not available.
§ Diffusion sample lost.

Table 3.3
Comparison of measured ^3H + tritiogenic ^3He to extrapolated ^3H input at Memphis and indication of dilution or mixing*

Sample ID (well number)	Hydrostratigraphic unit	Top of screen (feet below land surface)	Age using neon (yr)	Recharge year	Tritiogenic $^3He + ^3H$ (TU)	Average ^3H load for recharge year	1 standard deviation ^3H load for recharge year	Dilution (mixing)
Fall 2002 results								
MS-17	Memphis aquifer	132	26	1976	14.6	21	3.9	yes
MS-18	Memphis aquifer	155	32	1970	23.2	73.8	26.2	yes
MS-16 (C-2)	Memphis aquifer	93	33	1969	33.5	94.2	39.2	yes
44 (UC-3)	Shallow aquifer	45	~0	2002	3.6	1.45	1.1	
MS-12	Memphis aquifer	67.5	8	1994	7.0	4.2	2.2	
MS-13 (UC-1)	Memphis aquifer	110	9	1993	8.3	7	3.8	
MS-14 (UC-2)	Memphis aquifer	144	20	1982	7.3	13.6	3.1	yes
Spring 2003 results								
MS-17	Memphis aquifer	132	21	1982	15.7	13.6	3.1	
MS-18	Memphis aquifer	155	23	1980	15.7	17.7	3.2	
MS-16 (C-2)	Memphis aquifer	93	25	1978	21.9	32.9	11.6	
44 (UC-3)	Shallow aquifer	45	2	2001	3.9	1.5	1.1	
MS-13 (UC-1)	Memphis aquifer	110	13	1990	11.8	7.1	2.7	
MS-14 (UC-2)	Memphis aquifer	144	12	1991	12.8	6	4.2	
Fall 2003 results								
MS-15 (C-1)	Memphis aquifer	137	24	1979	21.1	16.4	4	
45 (C-3)	Confining unit	58	>40		0.1			
46 (C-4)	Shallow aquifer	31.5	9	1994	5.7	4.2	2.2	
Patriot Lake well	Memphis aquifer	200+	23	1980	2.2	17.7	3.2	yes

* Source of tritium input function data for comparison was obtained from Ivey (2003).

activity in rainfall during the year, measured tritiogenic helium+^{3}H values within one standard deviation of the expected ^{3}H loadings are considered to be balanced. If the measured tritiogenic helium+^{3}H is less than the expected ^{3}H loading by more than one standard deviation, then mixing with older water is suspected. If the measured tritiogenic helium+^{3}H is greater than the expected ^{3}H loading by more than one standard deviation, then either another source of ^{3}H is likely present or a mixture with more tritiogenic water is suspected. In the shallow aquifer wells and wells MS-12 and MS-13 at the unconfined site, the measured ^{3}H sources balance or slightly exceed the expected ^{3}H loading, suggesting that the ages are reasonable and that little or no mixing is occurring along the flow path.

In the deeper wells sampled during the Fall 2002, the measured ^{3}H sources are significantly less than the expected ^{3}H loadings, suggesting that mixing with older water may be occurring. However, results from the same wells during the Spring 2003 sampling event show balance between measured ^{3}H sources and expected ^{3}H loading. Although these data seem to conflict, they are consistent with the age data and seasonality of precipitation and groundwater pumping in the Memphis area. Monthly precipitation is greater during the spring months and municipal pumping rates are lower than those during the fall. The age results for water in the deeper wells (MS-14, -16, -17, and -18) in the Memphis aquifer from the Spring 2003 sampling event are all significantly (assuming 10% error) younger than those of the fall. Thus, the data suggest that recharge rates are greater during the spring months, resulting in overall younger water in the upper part of the Memphis aquifer and less mixing with deeper, older Memphis aquifer waters. The Patriot Lake well has a deeper screen depth than any of the other Memphis aquifer project wells and is up-gradient of the identified window area. The relatively low ^{3}H activity and sum of tritiogenic sources is consistent with water deeper within the Memphis aquifer, such as that sampled by studies in the Sheahan well field (Larsen et al., 2003; Ivey, 2003). The results from well MS-15 are difficult to interpret because this well produces little water, and thus was difficult to develop.

KRYPTON-85 STATUS

Although the development of the ^{85}Kr analysis method at the Institute for Rare Isotope Measurements (IRIM) at the University of Tennessee had not yet been completed when this research project was proposed, it was decided to include sampling and analysis for this emerging environmental tracer because of the opportunity to compare results with more established tracers, such as ^{3}H/^{3}He, in a setting that is particularly applicable to ^{85}Kr. Results from ^{85}Kr should be very similar to ^{3}H/^{3}He, as they both cover the same 0-40 year age-dating time period, but are perhaps a little easier to interpret in complex environments as the measurement is an isotope ratio of a single element, eliminating the possibility of differing chemistries. Unfortunately, the analytical method is technically very challenging, and it has become clear that no ^{85}Kr sample results will be available in time for this final project report. As the ^{3}H/^{3}He results permit us to complete the scientific goals of the project, it was decided not to further delay the project by waiting for the ^{85}Kr analysis method to come on-line.

The challenge of measuring ^{85}Kr from environmental samples is two-fold: first, krypton in itself is a rare constituent (1.1 ppm) of the earth's atmosphere and second, the isotopic abundance of ^{85}Kr is only one part in 10^{12}, so that one liter of modern water contains only 20,000 atoms (10^{-15} cc) of ^{85}Kr. A few ^{85}Kr measurements from groundwater have been made using low-level radioactive decay counting, but these required 200-liter water samples and only the group at the University of Bern (Switzerland) is capable of making them. The Bern lab is not currently active in

^{85}Kr research, so this was not an option for our samples. The analytical method for ^{85}Kr under development at IRIM can make these measurements from 3 to 10-liter samples, and consists of six major steps: (1) water sample collection without contamination by atmospheric gases, (2) degassing the water, (3) efficient separation of krypton from other gases, (4) isotopic enrichment in a high-throughput custom mass spectrometer, (5) further isotopic enrichment in a custom quadrupole mass spectrometer, and (6) counting of the few thousand remaining ^{85}Kr analyte atoms in a laser-based resonance ionization time-of-flight mass spectrometer with unprecedented sensitivity.

At the time this project was proposed, the development of the ^{85}Kr analytical method at IRIM was primarily sponsored by the National Science Foundation (NSF) and the U.S. Department of Energy (DOE). This funding lapsed for almost two years in the 2000 to 2001 time period, and IRIM had to drastically curtail its activities, losing highly experienced staff. Although NSF and DOE funding resumed in late 2001, progress has been slower than anticipated, primarily due to difficulties in staffing for this highly skilled and diverse laboratory environment. In spite of the above setbacks, significant progress has been made.

In step (1) - hermetic groundwater collection - a custom purging and pumping system was designed and constructed to meet the special requirements of this project, and six additional 20-liter ^{85}Kr sampling containers were constructed to keep delays in sample analyses from interfering with sample collection (Figure 3.1b).

Although step (2) - water degassing - was completely operational at the start of this project, many tests were run using "standardized" water samples, refining operating protocols to improve degassing efficiency and reproducibility, and assuring that no contamination from ambient air was occurring. It was these tests, repeated after the system had been idle for a year after processing the first round of samples that made us realize that "real" groundwater can be chemically rather active, and it was discovered that micro-cracks due to stress corrosion had developed in the stainless steel degassing chamber. However, no samples were compromised.

Step (3) - efficient separation of the 0.0015 cc of krypton from the typically 450 cc of air extracted from 20 liters of groundwater provided another surprise when it was discovered that some "real" groundwater samples contained much higher levels of CO_2 than "standardized" water samples, which were made from tap water that had been equilibrated with laboratory air. This caused freeze-up of traps in the separator, and considerable experimentation was required in solving this problem without contaminating or loosing the krypton from the sample.

Step (4) - first isotopic enrichment - utilizes a custom gas recirculating mass spectrometer with a plasma ion source coupled to a velocity filter to reduce the stable krypton isotopes by five orders of magnitude relative to ^{85}Kr. Although the krypton beam is intense, (100 nA) it still takes 10 hours to process 50% of the sample, and during this time period, argon, which is 1% in air, builds-up in the system. As the argon ions take a very different trajectory than the krypton ions, they strike the walls of the vacuum chamber, sputtering metal that causes internal short circuits after only a few enrichments. Repair costs several months of down-time. A solution has been designed and implementation will commence shortly.

Step (5) - second isotopic enrichment - utilizes a quadrupole mass spectrometer in a non-standard operating mode that has plagued us with variable memory and sample loss problems for a long time, precluding accurate results and reliable operation. During this project we implemented internally heated poles (to 500°C) that required significant redesign and construction, solving the memory problem. The sample-loss problem should improve by operating at a higher radio frequency. Equipment for this has been received and modifications to the system are almost completed.

Table 3.4
Water samples collected from the Shelby Farms study area for ^{85}Kr analysis

Collection date	Well designation	Quantity	Status
Nov. 14, 2002	MS-13	20 liters	DG
Nov. 14, 2002	MS-14	20 liters	DG
Nov. 14, 2002	MS-12	20 liters	DG
Nov. 14, 2002	44	20 liters	DG
Nov. 15, 2002	MS-16	20 liters	DG
Nov. 15, 2002	MS-17	20 liters	DG
Nov. 15, 2002	MS-18	20 liters	DG
June 9, 2003	MS-14	20 liters	AW
June 9, 2003	MS-13	20 liters	AW
June 9, 2003	MS-12	20 liters	AW
June 9, 2003	44	20 liters	AW
June 10, 2003	MS-18	20 liters	AW
June 10, 2003	MS-16	20 liters	AW
June 10, 2003	MS-17	20 liters	AW
Dec. 3, 2003	MS-15	20 liters	AW
Dec. 4, 2003	Patriot Lake	20 liters	AW

AW = archived as water in hermetic glass sampling bottle.
DG = degassed and stored in 4.8-liter hermetic aluminum cylinder.

Step 6 - counting the few thousand remaining ^{85}Kr atoms - utilizes a unique mass spectrometer with a pulsed sample concentrator and laser resonance ionization. The system has operated at full sensitivity for a number of years but has had to address a recurring problem that has finally been solved. A number of episodes of laser damage to optics were tracked down to a design failure in the main pump laser. Unfortunately, any repair to the main pump laser misaligns all of the downstream beams, requiring many days of tedious and delicate realignment. The offending components in the pump laser have been replaced with parts from a different manufacturer and have worked fine since then.

The samples collected during this project are safe and will be archived at IRIM. Processing of these samples will only resume when we are convinced that reliable ^{85}Kr results will be possible. A summary of the status of water samples collected for ^{85}Kr analysis is provided in Table 3.4.

GEOCHEMISTRY

Several types of field and laboratory analyses were conducted on samples collected from the Shelby Farms site. Temperature, pH, redox potential, and concentrations of dissolved oxygen, nitrate, and ammonia, were measured in the field at the time of sample collection using a flowthrough sampling cell and a calibrated YSI 6600 multiprobe unit. Alkalinity also was measured in duplicate with a Hach digital titration system and total dissolved iron and nitrate concentrations were measured using Hach ampules and a Hach 2400 spectrophotometer. During the December 2003 sampling event, dissolved sulfide concentrations were measured in the field using Hach reagents and a Hach 2400 spectrophotometer. Upon return of samples to the lab, anion concentrations were determined using external calibration standards and a Dionex 120 ion chromatography unit. Detection limits generally were 0.01 mg/L and one standard deviation errors generally were 10% or less for the anion concentrations observed in the Shelby Farms samples. Lab blanks analyzed in association with the samples generally had anion concentrations less than detection limits, but values as high as 0.05 mg/L were occasionally observed for chloride. Field equipment blanks generally had detectable anion concentrations with chloride concentrations approaching 0.2 mg/L. Cation concentrations were determined in the laboratory using matrix-matched external calibration standards and a Varian FS220 atomic absorption spectrometer. Detection limits generally were 0.01 mg/L or less and one standard deviation errors generally were 5% for all analyzed metals except iron, which had an error of 8%. Lab blanks generally had concentrations of cations less than 0.04 mg/L. Field equipment-blank values range from less than detection for iron to 0.6 mg/L for sodium. Duplicate sample analyses had errors of 5% or less. Charge balances for each total sample analysis generally were within 10% of electrical neutrality. Cation determinations for the December 2003 event samples and all trace element analyses were determined by inductively coupled plasma–optical emission spectroscopy at the University of Mississippi. The common cations had errors (% root standard deviation) of 6% or less and the trace elements had errors of 11% or less.

Chemical analysis of waters collected from monitoring wells in November 2002 and December 2003 are presented in Table 3.5. A trilinear diagram illustrating the major- ion composition of waters sampled in 2002 and 2003 is presented in Figure 3.2. Another trilinear diagram illustrating the major-ion composition of all waters sampled at Shelby Farms is presented in Figure 3.3. The chemistry of the Patriot Lake well is similar to that of background water chemistry described by Parks and Mirecki (1992) and is interpreted to represent background water chemistry in the Memphis aquifer at the site. The trilinear diagrams illustrate that a range of water compositions exist at the Shelby Farms site. Possible end-member compositions include Ca-Mg-Na-SO_4, mixed cation-Cl, and Ca-Na-HCO_3 waters.

Most of the other sampled shallow groundwaters at the Shelby Farms site are variably influenced by landfill leachate, similar to observations from previous studies (Bradley, 1991; Parks and Mirecki, 1992). Mixing with leachate generally increases the total dissolved solids (Figure 3.4), decreases the redox state of the water, and increases the concentrations of several dissolved constituents, such as barium, boron, calcium, chloride, potassium, sodium, and strontium (Mirecki and Parks, 1994). Mirecki and Parks (1994) also showed that barium and strontium were useful tracers for determining the percentage of leachate plume water mixed with the native waters. Table 3.6 shows the mixing proportions as determined by linear mixing equations presented in Mirecki and Parks (1994). The mixing proportions generally follow the qualitative trends shown in Figures 3.4 through 3.8; however, the highest mixing proportions in Table 3.6 are

Table 3.5
Field and laboratory chemical analysis and $^3H/^3He$ data of water sampled in November 2002 and December 2003

Sample ID (well number)	$^3H/^3He$ data			Field data						
	Hydrostratigraphic unit	TritHe^3 using Ne (TU)	Tritium (TU)	Age using Ne (yr)	Top of screen (ft)	Top of screen (m)	Temp. (°C)	Cond. (mS/cm)	pH	Alkalinity (mg/L $CaCO_3$)
2002 results										
MS-17	Memphis	11.2	3.4	26	132	40	17.08	0.299	6.19	111
MS-18	Memphis	19.2	3.9	32	155	47	17.3	0.289	6.2	68.2
MS-16 (C-2)	Memphis	28.2	5.4	33	93	28	17.46	0.242	6.12	53.2
44 (UC-3)	Shallow	-0.4	3.6	nd	45	14	17.9	0.256	5.74	90.4
MS-12	Memphis	2.5	4.5	8	67.5	21	18.26	0.484	5.98	200
MS-13 (UC-1)	Memphis	3.2	5.1	9	110	34	19.28	0.431	6.25	276
MS-14 (UC-2)	Memphis	4.9	2.4	20	144	44	18.79	0.962	6.32	428
2003 results										
MS-15 (C-1)	Memphis	15.5	5.6	24	137	42	14.19	0.41	6.01	150
45 (C-3)	Confining unit	-2.9	0.1	>40	58	18	12.8	nd	6.29	138
46 (C-4)	Shallow	2.3	3.5	9	31.5	10	15.67	0.134	5.56	37
Patriot Lake well	Memphis	1.6	0.6	23	290	88	19.51	0.157	5.94	42

(continued)

Table 3.5 (Continued)

	Field data					Laboratory analysis					
Sample ID	Iron (total) (mg/L)	Dis. O_2 (mg/L)	Ammonium (mg/L)	Sulfide (μg/L)	Platinum probe redox (mV)	NO_3^-/NH_4^+ redox (V)	SO_4^{2-}/S_2^- redox (V)	Fluoride (mg/L)	Chloride (mg/L)	Nitrate (mg/L)	Sulfate (mg/L)
2002 results											
MS-17	2.1	0.32	0.12	nd	275			0.14	7.18	0	26.8
MS-18	0.05	2.66	0.09	nd	389	0.45		0.1	13.9	26.6	24.9
MS-16 (C-2)	1.86	0.23	0.08	nd	44			0.08	3.98	0	55.6
44 (UC-3)	nd	2.06	0.06	nd	356	0.49		0.11	1.41	14.5	27.2
MS-12	3.3	0.19	0.10	nd	183			0.27	2.88	0	46.3
MS-13 (UC-1)	3.3	0.92	3.4	nd	244			0.32	28.7	0	4.67
MS-14 (UC-2)	3.3	0.49	10.5	nd	192			0.5	38	0	0
2003 results											
MS-15 (C-1)	nd	0.132	0.04	83	nd	0.45	-0.11	0.07	4.4	0.007	75
45 (C-3)	nd	4.05	0.2	463	nd	0.43	-0.13	0.12	1.7	0.04	33.3
46 (C-4)	nd	4.4	0.02	16	nd	0.50	-0.08	0.006	4.8	3.8	14.8
Patriot Lake well	nd	0	0.02	18	nd		-0.11	0.04	8.9	0	13.5

(continued)

Table 3.5 (Continued)

	Laboratory analysis										
Sample ID	Fe (mg/L)	Mn (mg/L)	Na (mg/L)	K (mg/L)	Ca (mg/L)	Mg (mg/L)	Ba (mg/L)	B (mg/L)	Sr (mg/L)	Total diss. solids (mg/L)	Elec. neut.
2002 results											
MS-17	1.8	0.3	21.6	0.8	27.0	9.3	0.08	0.09	0.12	206	5.64
MS-18	0.0	0.2	38.4	1.1	13.5	5.3	0.08	0.09	0.07	192	3.95
MS-16 (C-2)	1.6	0.1	11.4	1.1	19.9	8.6	0.09	0.3	0.07	156	-2.05
44 (UC-3)	0.0	0.2	39.5	0.8	11.8	5.4	nd	nd	nd	191	4.74
MS-12	5.5	0.7	56.7	1.0	32.1	13.2	0.07	0.09	0.09	359	6.87
MS-13 (UC-1)	69.0	0.4	29.0	4.2	26.1	14.3	0.12	0.15	0.11	456	1.41
MS-14 (UC-2)	76.8	0.4	37.5	16.7	60.4	23.0	0.21	0.29	0.26	692	6.51
2003 results											
MS-15 (C-1)	43.51	0.79	22.8	1.56	26.48	9.45	0.2	0.1	0.12	335	0.45
45 (C-3)	19.67	0.92	11.72	4.6	27.12	12.75	0.4	0.13	0.3	251	6.94
46 (C-4)	0.12	0	8.89	1.32	11.03	4.73	0.05	0.11	0.03	87	9.21
Patriot Lake well	1.03	0	10.86	0	14.14	4.42	0	0.08	0.02	95	13.83

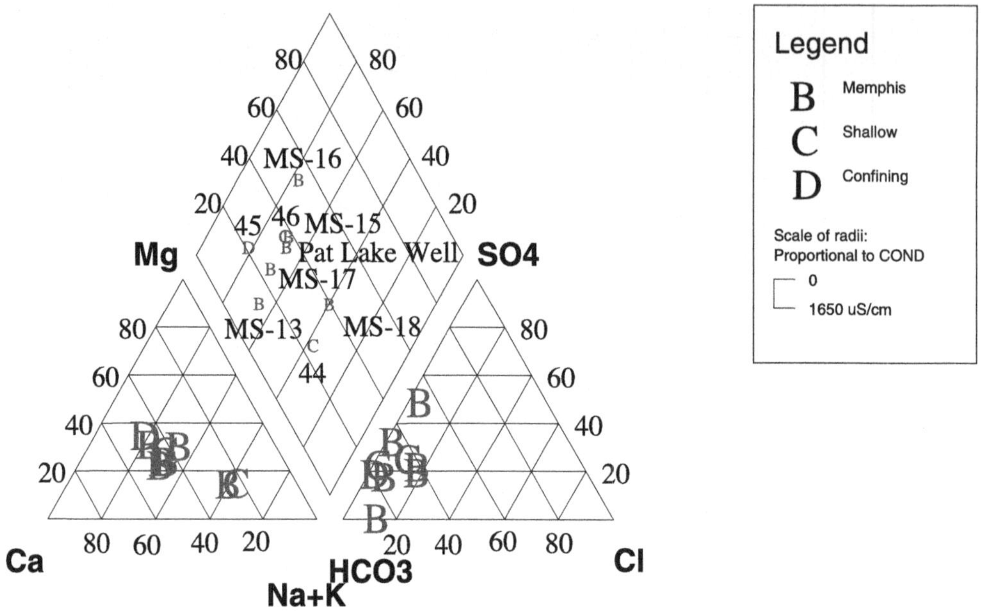

Figure 3.2 Trilinear plot of data from November 2002 and December 2003 sampling events

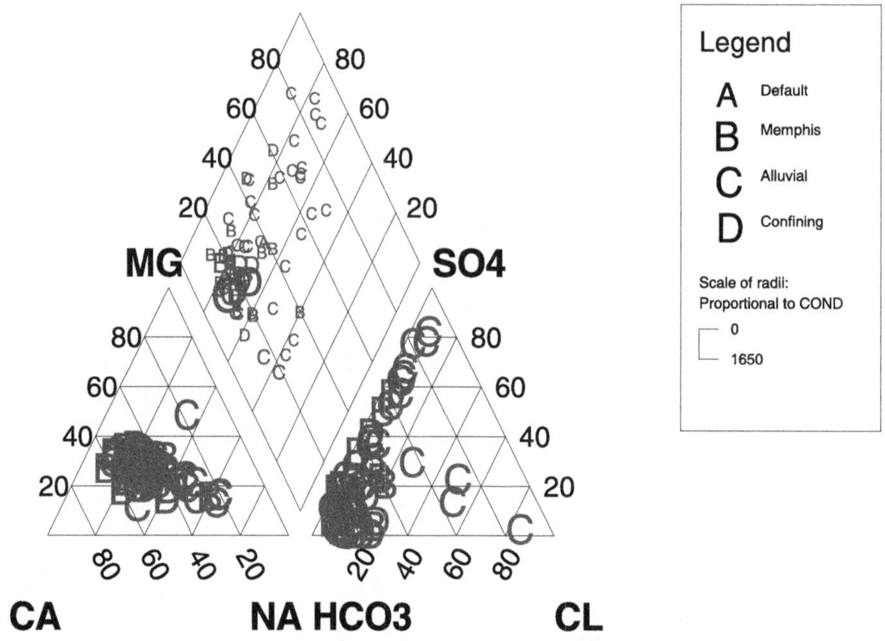

Figure 3.3 Trilinear plot of all available data from wells at the Shelby Farms site

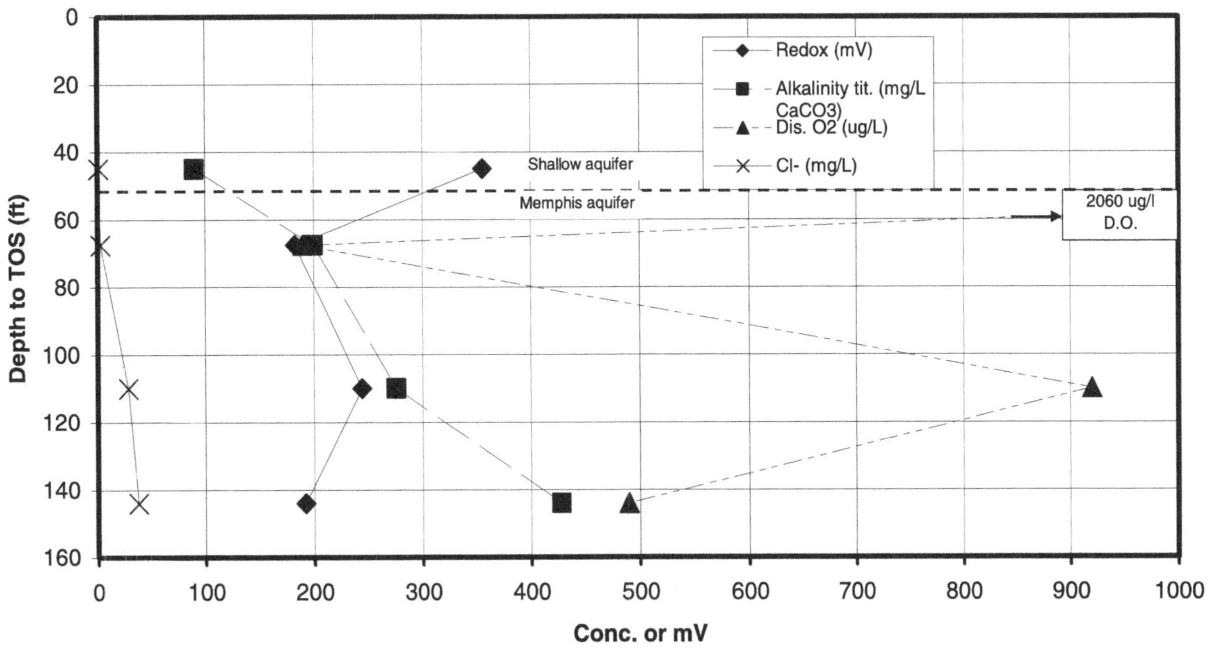

Figure 3.4 Plot of selected field and lab data versus depth for the unconfined cluster November 2002

those from the confining unit well (45), suggesting that barium and strontium have non-contaminant sources in the clay.

The most chloride-rich end-member waters are from the shallow aquifer and confining unit (Figure 3.5), but some of the Memphis aquifer wells sampled in 2002 and 2003 have elevated chloride levels as well. Mirecki and Parks (1994) indicated that while chloride does show a relationship to leachate influence it is not as consistent as that of barium and strontium.

The results from the unconfined cluster show a systematic change in redox buffers and overall decreasing redox state with depth, typical of waters influenced by landfill leachate (Baedecker and Back, 1979a, b; Lyngkilde and Christensen, 1992). The geochemical patterns at the confined cluster show more complex relationships with depth: uncontaminated water in the shallow aquifer well, chemically evolved water in the confining unit well (that bears resemblance to leachate-influenced waters), and leachate influenced wells in the Memphis aquifer (Figures 3.7 and 3.8). Other wells downgradient from the window, MS-17 and MS-18, also show some indications of leachate influence (high nitrate and sulfate, depressed redox, etc.), but are likely to be at the margins of the leachate-influenced area. Although the chemistry of water from the confining unit well (45) at the confined cluster appears to be similar to that of the leachate-influenced waters, it contains no modern water and is likely to reflect long-term water-rock interaction in the clay-rich and pyrite- and siderite-bearing confining unit clay.

Geochemical data were entered into PHREEQCI, a geochemical modeling program, to evaluate redox-pair (e.g., Sulfate-sulfide, etc.) Eh conditions and mineral saturation. Because slightly different field analyses were completed during the 2002 and 2003 sampling events, the modeling results vary slightly. Most waters sampled are well out of redox equilibrium, which is not unusual. Sulfur, nitrogen, and dissolved oxygen couples yield differing Eh values, although

Table 3.6
Percentage of leachate in sampled waters as determined using linear mixing relationships developed by Mirecki and Parks

	Ba mg/L	% leachate using Mirecki and Parks (1994)	Sr mg/L	% leachate using Mirecki and Parks (1994)
2002 results				
MS-17	0.08	6	0.12	21
MS-18	0.08	6	0.07	6
MS-16	0.09	8	0.07	6
44	nd*		nd	
MS-12	0.07	3	0.09	12
MS-13	0.12	16	0.11	18
MS-14	0.21	38	0.26	63
2003 results				
MS-15	0.2	36	0.12	21
45	0.4	86	0.3	75
46	0.05	<0	0.03	<0
Patriot Lake well	Bd	<0	0.02	<0

* nd indicates not determined for that sample.

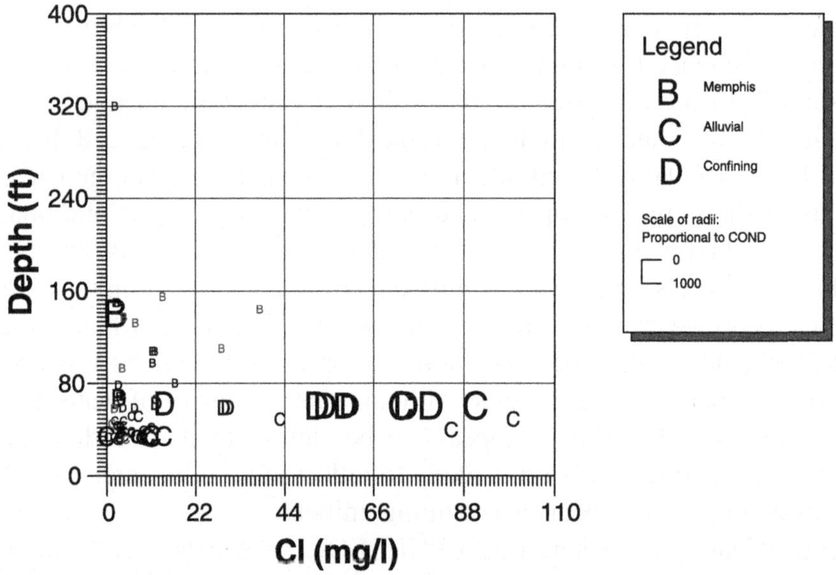

Figure 3.5 Cross-plot of chloride versus depth for all data at Shelby Farms site. Data from 2002-2003 sampling events are highlighted.

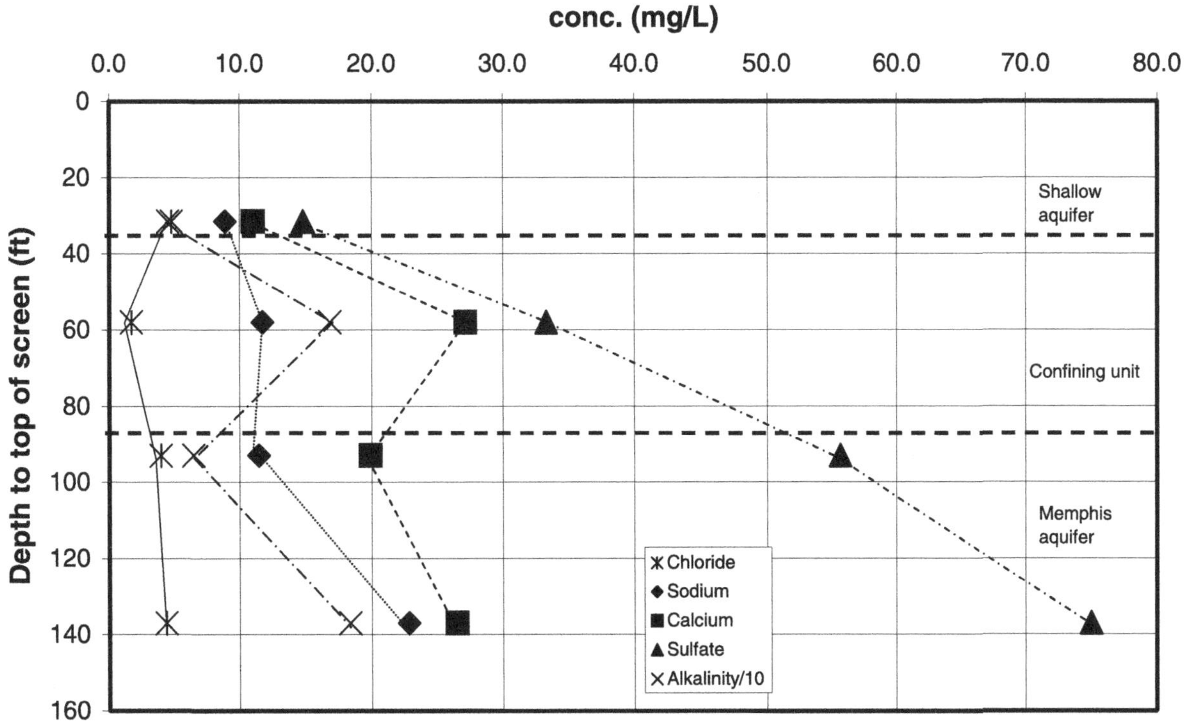

Figure 3.6 Selected major dissolved species in water sampled at the confined cluster. Data from both November 2002 and December 2003 are shown.

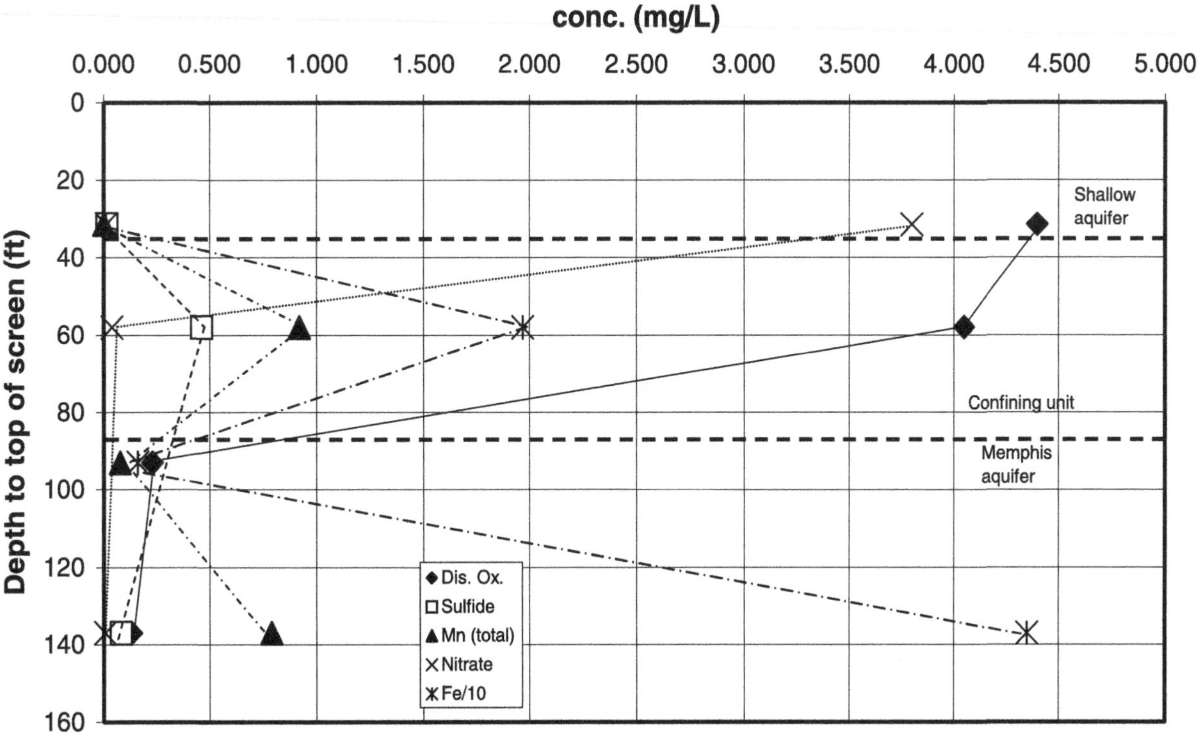

Figure 3.7 Selected redox-sensitive species in water sampled at the confined cluster. Data from both November 2002 and December 2003 are shown.

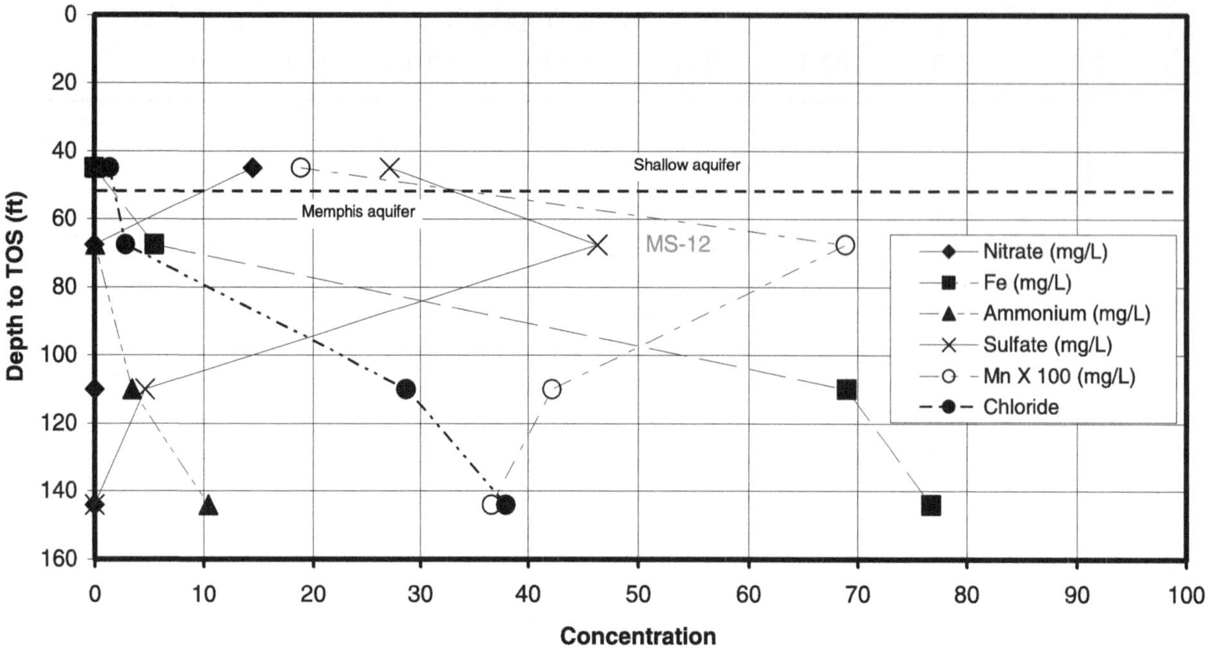

Figure 3.8 Plot of selected redox data versus depth for the unconfined cluster, November 2002

usually the concentration of one redox couple dominates over the others suggesting that redox buffering is occurring (as is illustrated in Figure 3.8).

In general, most of the sampled waters are oversaturated with various iron oxides and oxyhydroxides, and typically approach equilibrium with amorphous ferric hydroxide. Sampled waters from the leachate-influenced wells are near equilibrium or oversaturated with respect to barite ($BaSO_4$), an observation also made by Mirecki and Parks (1994). Most of the reduced, sulfide-bearing waters are oversaturated with respect to pyrite and are near equilibrium with siderite. The modeling results suggest that some mineral reactions, especially those involving iron, carbonate, sulfate-sulfide, and barium, may be important controls on groundwater chemistry at the Shelby Farms site.

CHAPTER 4
GROUNDWATER FLOW SIMULATION

MODELING PURPOSE AND OBJECTIVES

The purpose of the modeling effort was to assimilate the hydrostratigraphic data and environmental tracer data and access the suitability of the conceptual model proposed for the site. As originally proposed, the modeling effort built upon an existing calibrated model (Ng, 1993) of the Shelby Farms area. The modeling analyses consisted of three primary objectives that contributed to the overall goals of the project. The objectives were: (1) using the unaltered calibrated transmissivities by Ng (1993), determine the most highly probable area of leakage to the Memphis aquifer at the site using an inverse genetic algorithm technique (Gentry et al., 2001); (2) use the inverse parameter estimation package (modified Gauss-Newton search technique) associated with MODFLOW 2000 (Harbaugh et al., 2000) to corroborate the genetic algorithm findings; and (3) use the environmental tracer data to corroborate or modify the flow model as needed.

MODEL DEVELOPMENT IN ARGUS OPEN NUMERICAL ENVIRONMENT

The original calibrated data from the historical Shelby Farm model (Ng, 1993) was exported and used in a steady-state finite-difference algorithm developed by Gentry et al. (2001). The same properties data was imported into Argus Open Numerical Environments (ArgusONE; Shapiro et al., 1997) and MODFLOW 2000 (Harbaugh et al., 2000). ArgusONE provides specific processing tools for MODFLOW 2000 that has been developed by USGS; ArgusONE was adopted for this project due to the ease of sharing of the model and data.

MODEL BOUNDARIES AND DISCRETIZATION

The extents and boundary conditions of the model were developed from models previously used to investigate leakage at the Shelby Farms study area and are shown in Figure 4.1 (Ng, 1993; Gentry et al., 2003). Previous plans had detailed the construction of a multilayer model for flow simulation. However, due to uncertainty involving the size of the leakage area, the boundary conditions associated with the alluvial aquifer, the lack of hydrologic control data in the vicinity for the deeper part of the Memphis aquifer, and the relatively short travel times demonstrated by the environmental tracer data, modeling of the shallow aquifer was considered unnecessary and a single layer representing the Memphis aquifer was constructed. Most of the data for the Memphis aquifer in the local study area is limited to the upper most portion of the aquifer with little data for the deeper sections of the formation. For the stated purpose of this research, the 2-D model application is valid and useful.

The boundary conditions and transmissivities were based upon historical investigations performed by the USGS (Bradley, 1990; Parks, 1990; Kingsbury and Parks, 1993; Brahana and Broshears, 2001). The modeled study area consists of a 20,000 ft by 12,500 ft rectangular area rotated 33° from the east-west horizontal plane along the principal axis of flow. Based upon historical data, the suspected extent of the window at the site is shown in Figure 4.1 (Ng, 1993). A uniform 50 by 80 cell grid was utilized to discretize the numerical representation of the study area. Each finite difference cell is 250 ft by 250 ft for a total of 4,000 cells or nodes. All nodes in

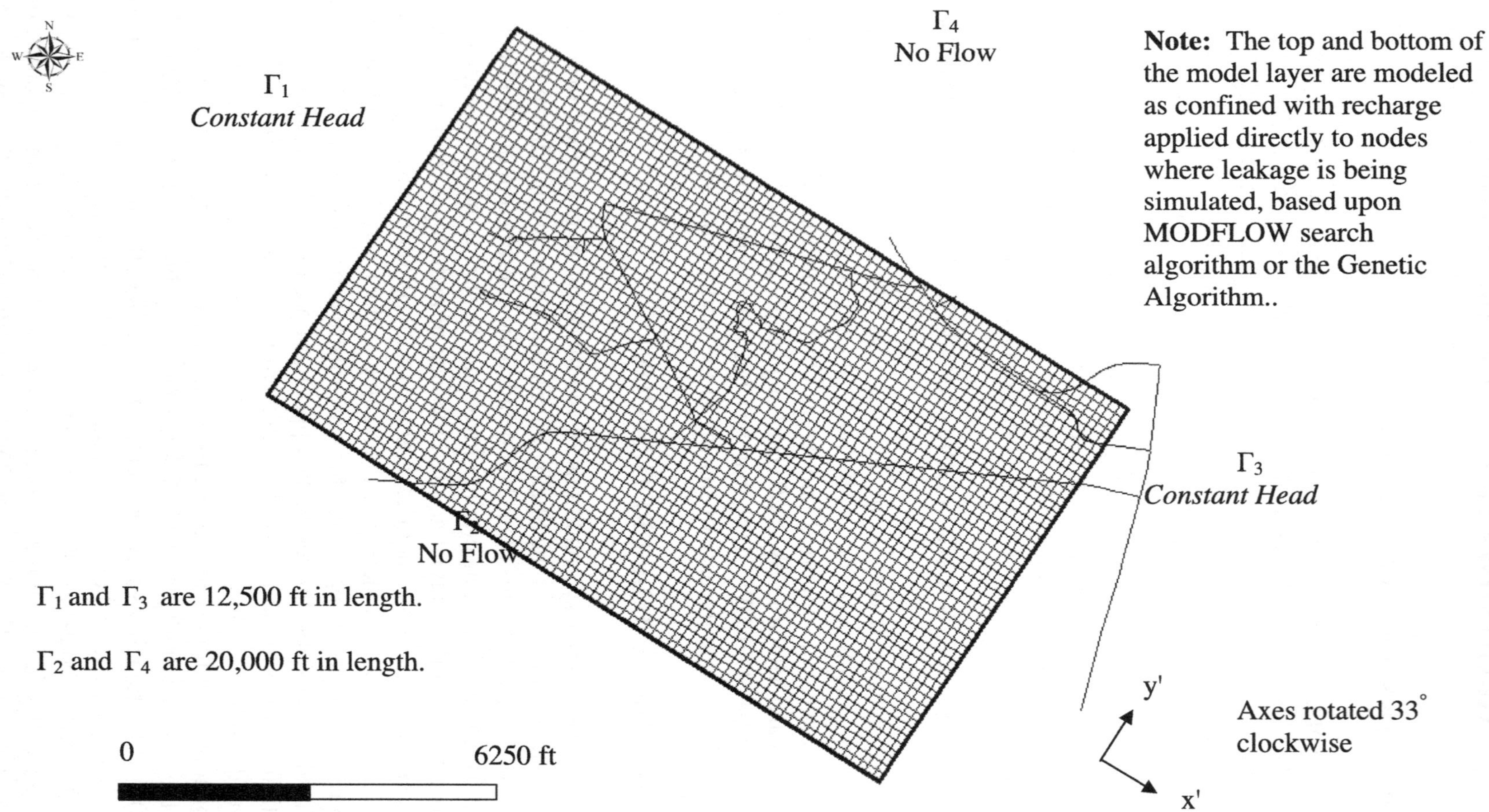

Figure 4.1 Shelby Farms numerical simulation model extents and boundary conditions for Memphis aquifer layer

the model mesh are active. The 1988 observation data and boundary conditions were used for the genetic algorithm evaluation. For subsequent analyses, more recently obtained observation (1999) data were used and the constant head boundary values were updated.

The behavior of the Memphis aquifer system was assumed to be similar to that of the 1988 conditions with the general flow path direction still be in the northwest direction (see Figure 2.15). Data for well MS-12 and Q-1 from Parks and Mirecki (1992) were compared with measurements taken in 2002. A decline in water level from the 1988 elevation of 4.78 feet at MS-12 and 5.55 feet at Q-1 were calculated. Therefore the constant head boundaries used in the previous model (Gentry et al., 2003) were adjusted by subtracting the average of the two points selected in the domain, 5.17 feet. The northwest and southeast faces of the model domain are defined as constant-head boundaries (Dirichlet conditions). The northeast and southwest faces are defined as no-flow boundaries (Neuman conditions). These boundary conditions are consistent with previous hydrologic investigations and published potentiometric maps of the study area (Bradley, 1991; Parks, 1990; Kingsbury and Parks, 1993). Also, the top and bottom of the layer are modeled as confined with recharge applied as a source term to specific cells directly.

INVERSE MODELING OF LEAKAGE

Gentry et al. (2003) used a genetic algorithm (GA) search technique coupled with a single layer central finite-difference model to identify the most probable area of leakage at the Shelby Farms study area, given the hydraulic-head data collected in 1987 (Bradley, 1988). The search area consisted of a rectangular area, 484 nodes, in the immediate vicinity of the observation wells and is shown in Figure 4.2. The GA used in this study is a modified version of a program originally developed by David Carroll at the University of Illinois (Carroll, 1997) and was modified to develop an inverse approach (Gentry et al., 2001). The source code for the GA driver is free for public use and is available over the Internet. Carroll's program is a FORTRAN version of a genetic algorithm driver. For this study, the GA used a binary mapping method to identify cells that were either "on," receiving accretion, or "off," not receiving accretion. Accretion rates utilized in the evaluation varied from 0.05 ft/day to 2 ft/day. The broad range of values was considered in order to ensure a robust search area for the variable and included the most likely range of values that represented the accretion rate.

For each accretion rate, the best achievable error rate was recorded and average probability regions were determined. For example, the analysis of the GA run with an accretion rate of 0.05 ft/day indicated that the best achievable error for the single evaluation was 3.5 ft. The best achievable error and associated accretion flux calculated for each accretion rate are summarized in Table 4.1 (Gentry et al., 2003). Analyses using accretion rates of 0.09 ft/day, 0.1 ft/day, and 0.4 ft/day are shown in Figure 4.3.

The results from multiple GA runs indicated that the lowest numerical error was associated with accretion rates ranging from 0.4 ft/day to 1.5 ft/day, see Table 4.1, however these leakage rates may not be physically feasible. Over the range of lowest error, the calculated flux ranged from 425×10^3 ft^3/day and 468×10^3 ft^3/day. It should be noted that the inverse modeling used for the previously mentioned GA technique used observation data and boundary conditions from 1988 data.

Once the updated boundary conditions were established, six zones were identified, consistent with the conceptual model, and selected for inverse parameter estimation in MODFLOW 2000. The zones are identified by location in Figure 4.4. The recharge parameter was log transformed so that only positive values of recharge in the identified zones would be allowed. The

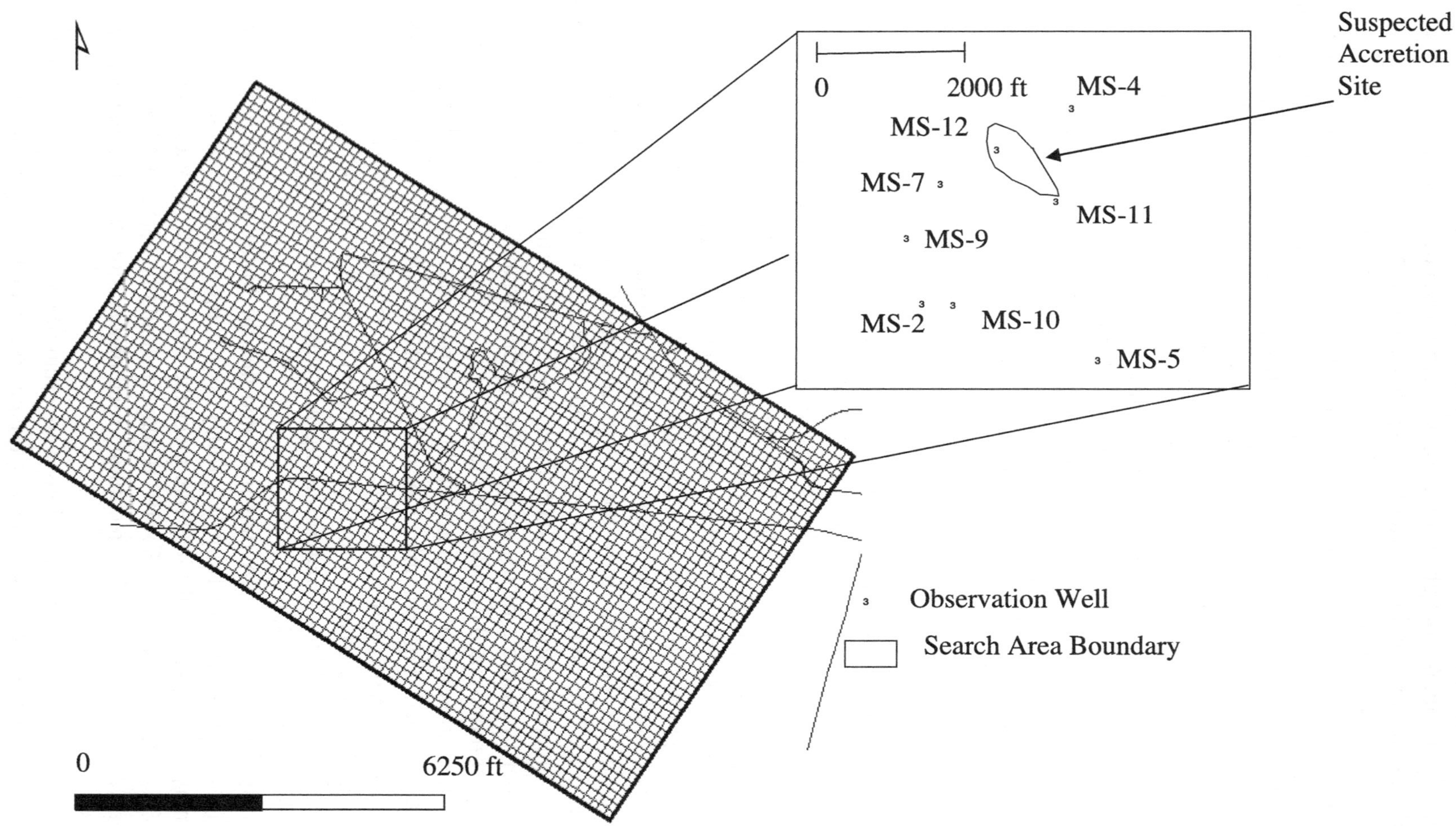

Figure 4.2 Genetic algorithm search area in the model domain

Table 4.1
The best achievable error for a genetic algorithm run

Accretion rate (ft/day)	Flux ($\times 10^3$ ft^3/day)	Error* (ft)
0.05	478	3.5
0.09	461	3.1
0.10	450	3.0
0.20	425	2.6
0.40	425	2.4
0.60	450	2.6
0.80	400	2.4
1.00	438	2.2
1.10	413	2.7
1.20	450	2.9
1.50	469	2.1
2.00	375	3.1
Average:	436	2.7
Standard Deviation:	29	0.36

Source: Adapted from Gentry et al., 2003.

* Error is expressed as, Error $= \sqrt{\sum (h_o - h_p)^2}$, for all GA runs.

modified Gauss-Newton method utilized by MODFLOW 2000 was allowed to iterate to the most probable values of recharge in the six zones.

For the first numerical experiment, the model was considered calibrated (without changing transmissivity values) with a total error of 13.62 feet, and is summarized in Table 4.2. A graph demonstrating the distribution of the error is shown as Figure 4.5. The final recharge values assigned to each zone identified in Figure 4.4 are summarized in Table 4.3. The steady-state head distribution and flow balance for the calibrated model is shown on Figure 4.6.

A second experiment was performed, whereby the transmissivity was adjusted globally in the model. After a series of initial trials, the transmissivity selected for demonstrating the effects of transmissivity reduction of the model results was 8% of the original. Again, the modified Gauss-Newton search technique used by MODFLOW 2000 was used to determine the most probable leakage in the six zones identified in Figure 4.4. For the second numerical experiment, the model was considered calibrated (adjusting the transmissivity values to within 8% of original) with a total error of 10.73 feet, and is summarized in Table 4.4. A graph demonstrating the distribution of the error is shown as Figure 4.7. The final recharge values assigned to each zone identified in Figure 4.4 are summarized in Table 4.5. The steady-state head distribution and flow balance for the calibrated model is shown on Figure 4.8.

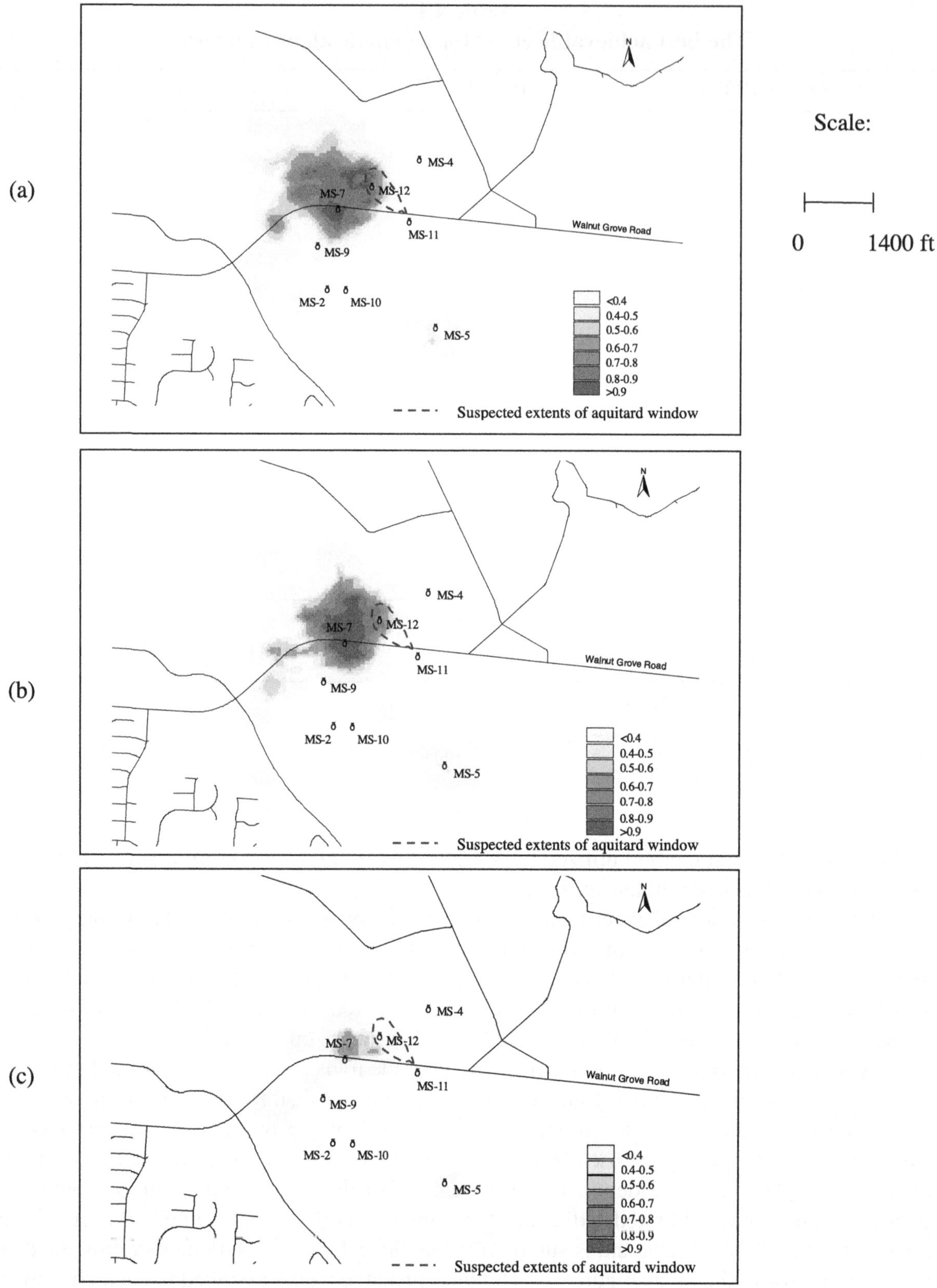

Source: Adapted from Gentry et al., 2003. Efficacy of Genetic Algorithm to Investigate Small Scale Aquitard Leakage. *ASCE Journal of Hydraulic Engineering, 12a(7):527-535*. With permission from ASCE.

Figure 4.3 Average probability regions using (a) R = 0.9 ft/day, (b) R = 0.01 ft/day, and (c) R = 0.4 ft/day

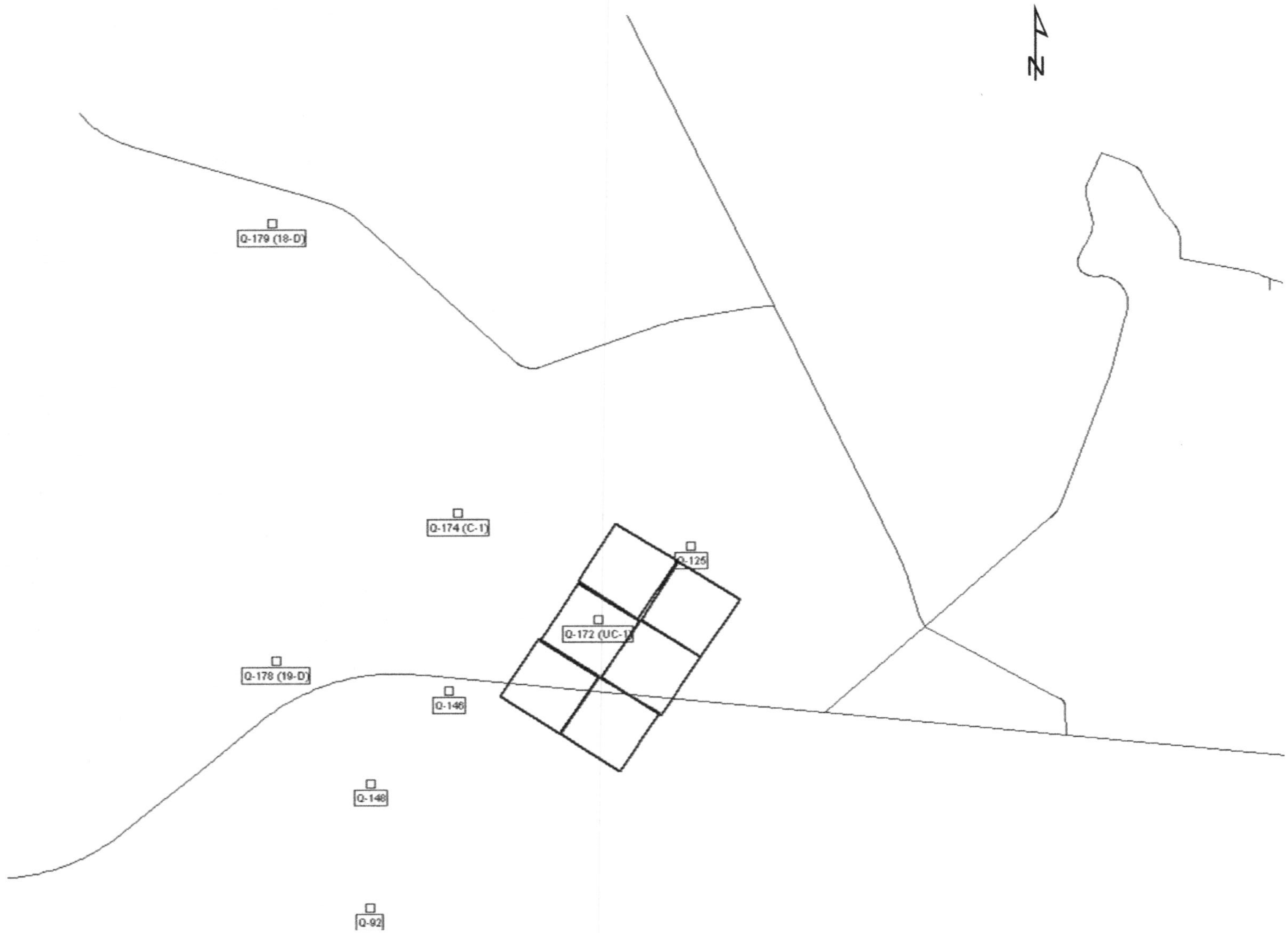

Figure 4.4 MODFLOW 2000 parameter estimation zones for area for Shelby Farms simulation model

Table 4.2
Calibrated model error for Experiment One (original transmissivities)

Location	Observed head(h_o) (NGVD ft)	Predicted head(h_o) (NGVD ft)	$(h_o - h_p)^2$ (ft^2)
Q-125	208.32	215.11	46.10
Q-92	211.15	212.06	0.83
Q-148	210.17	212.26	4.36
Q-179	202.30	200.73	2.46
Q-178	204.54	208.49	15.59
Q-174	206.45	211.83	28.89
Q-172	208.65	217.91	85.82
Q-146	216.72	215.50	1.50

$\sqrt{(h_o - h_p)^2} = 13.62$ ft

h_o = observed head

h_p = predicted head

NGVD - National Geodetic Vertical Datum of 1929.

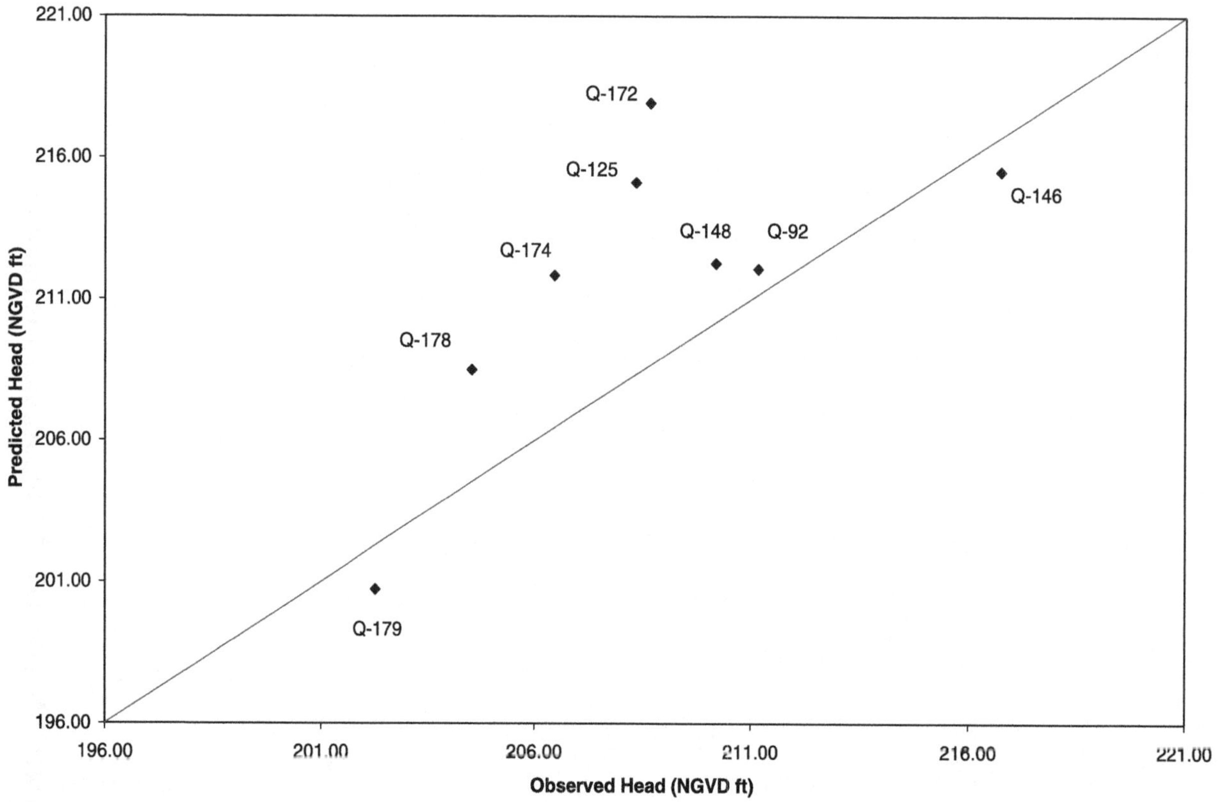

Figure 4.5 Graph of the error distribution for the Experiment One calibrated simulation model

Table 4.3
Recharge rates determined for Experiment One calibrated simulation model

	Recharge rate (ft/d)	Discharge (ft^3/d)
10	0.661	165346.70
20	0.000	0.00
30	0.106	26500.32
40	1.930	482802.47
50	0.000	0.00
60	0.000	0.00
	Sum:	674649.49 ft^3/d

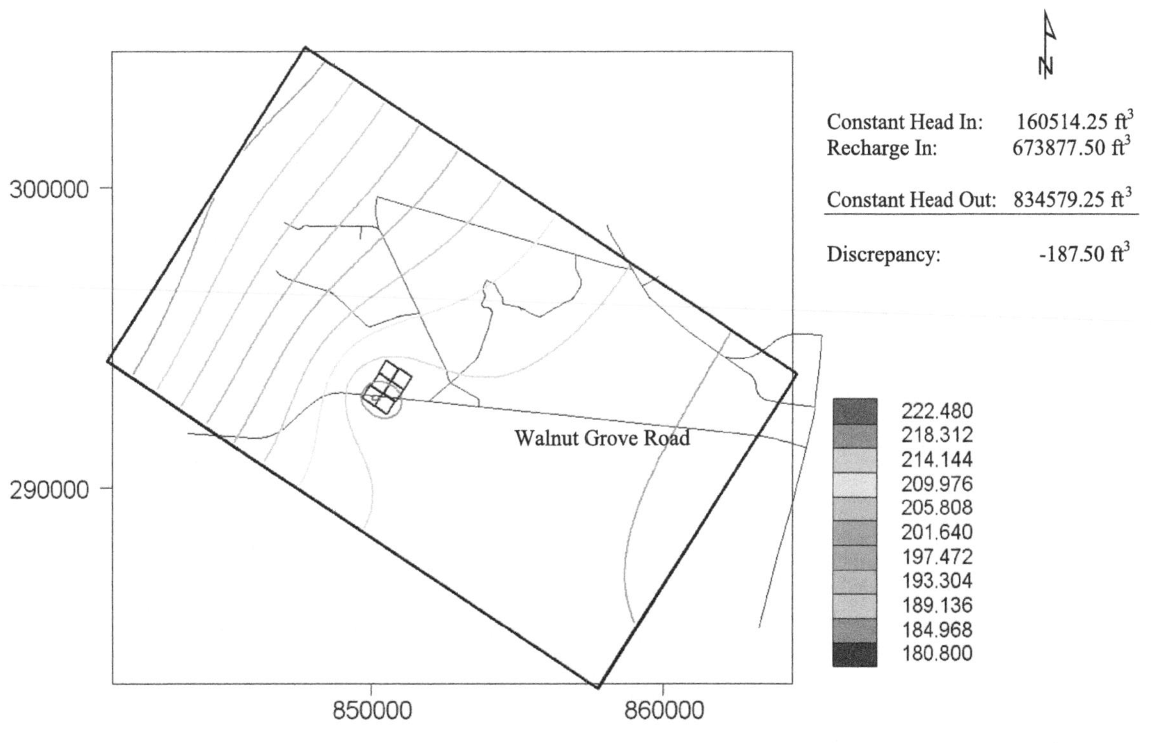

Figure 4.6 Steady-state head distribution and flow balance for Experiment One calibrated model

Table 4.4
Calibrated model error for Experiment Two (transmissivity globally adjusted)

Location	Observed head(h_o) (NGVD ft)	Predicted head(h_o) (NGVD ft)	$(h_o - h_p)^2$ (ft^2)
Q-125	208.32	209.76	2.06
Q-92	211.15	207.08	16.59
Q-148	210.17	206.90	10.69
Q-179	202.30	197.81	20.21
Q-178	204.54	203.82	0.52
Q-174	206.45	206.49	0.00
Q-172	208.65	211.21	6.56
Q-146	216.72	209.07	58.49

$\sqrt{(h_o - h_p)^2} = 10.73$ ft
h_o = observed head
h_p = predicted head

NGVD - National Geodetic Vertical Datum of 1929.

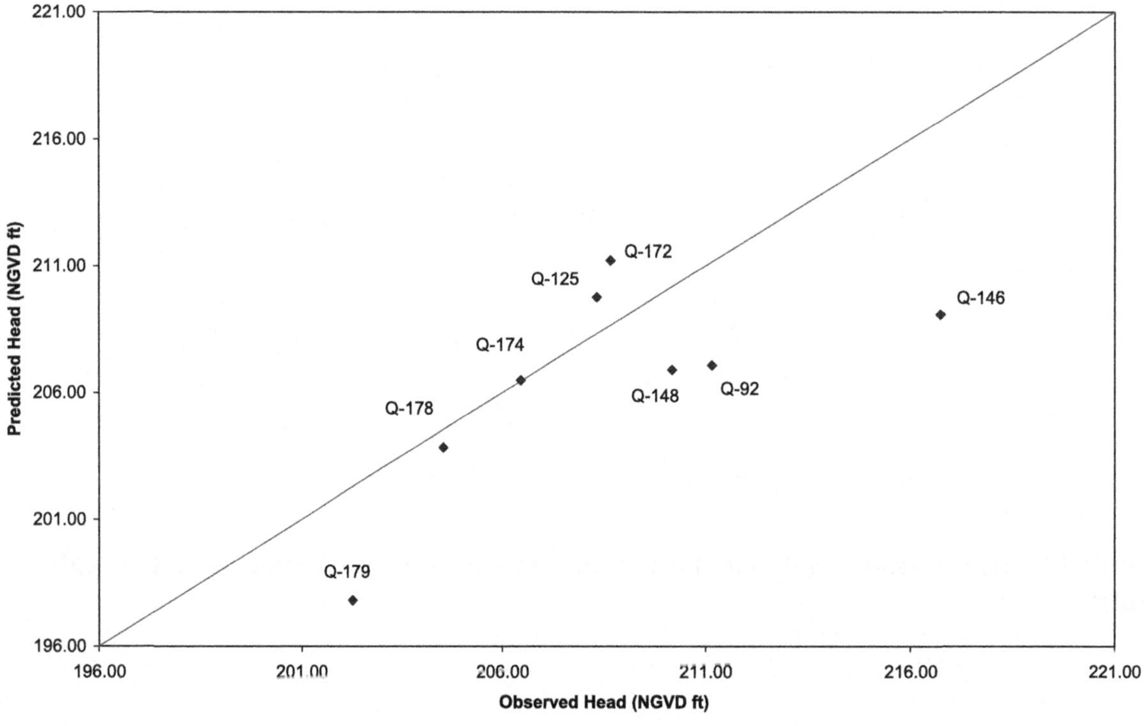

Figure 4.7 Graph of the error distribution for the Experiment Two calibrated simulation model

Table 4.5
Recharge rates determined for Experiment Two calibrated simulation model

Zone	Recharge rate (ft/d)	Discharge (ft^3/d)
10	3.44E-02	8614.36
20	5.53E-03	1385.31
30	1.09E-02	2731.38
40	9.00E-02	22525.78
50	9.74E-05	24.37
60	1.54E-03	386.44
	Sum:	35667.63 ft^3/d

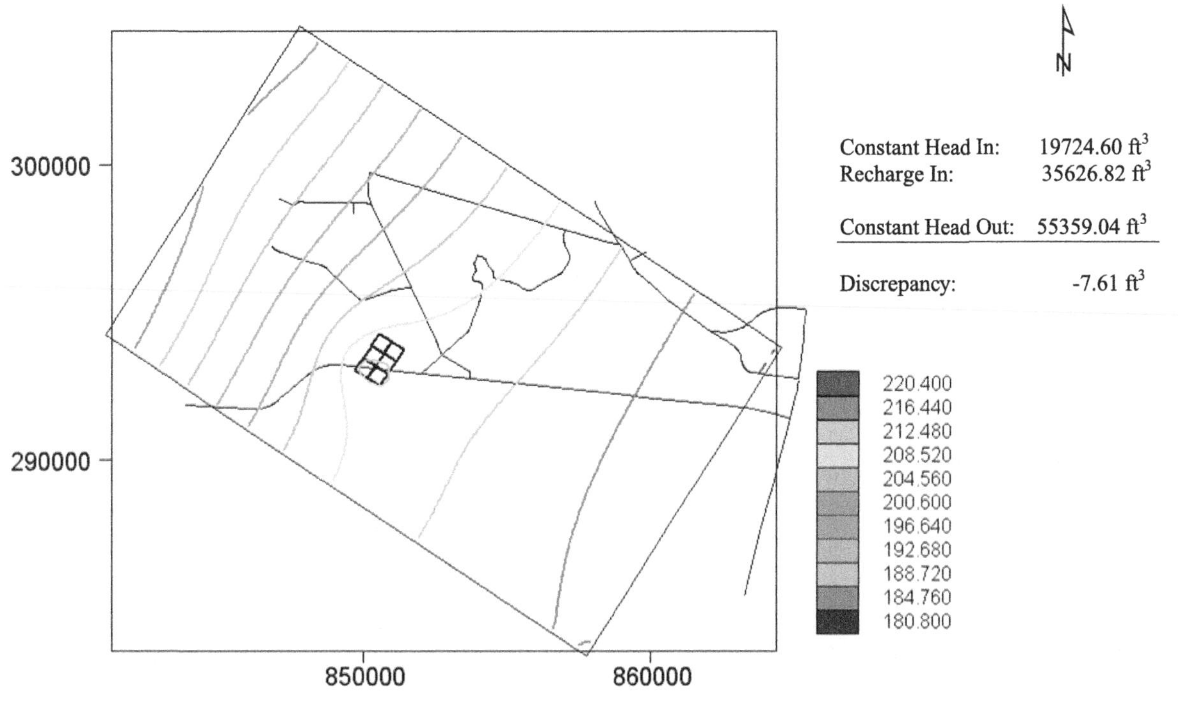

Figure 4.8 Steady-state head distribution and flow balance for Experiment Two calibrated model

The genetic algorithm modeling predicted flux (425×10^3 ft^3/day to 468×10^3 ft^3/day) and MODFLOW 2000 estimated flux for Experiment One (674×10^3 ft^3/day) were consistent. The MODFLOW 2000 estimated flux for Experiment Two (36×10^3 ft^3) is an order of magnitude lower. In order to affect the change in head necessary given the full transmissivity of the Memphis aquifer, a high recharge flux is necessary as demonstrated by the genetic algorithm and MODFLOW 2000 Experiment One results. If the aquifer behaves such that the full transmissivity is not available for flow then a much lower flux can be established that would affect the necessary head values. These concepts will be explored further in Chapter 5.

CHAPTER 5
DISCUSSION AND CONCLUSIONS

WELL HYDRAULIC TEST K's VERSUS GRAIN-SIZE K's

Comparison of hydraulic-conductivity values estimated from slug and pumping-tests of wells completed in the Memphis aquifer with estimates from the grain-size analyses of core samples collected from this unit shows considerable differences between the two data sets. Estimates of hydraulic conductivity based on grain-size results range from 88 to 156 ft/d (Figures 2.12 and 2.13) for samples collected from the Memphis aquifer at both the confined and unconfined well-cluster locations, with an average of about 116 ft/d for both sites. Values determined from the grain-size analyses are representative of disturbed samples that have lost any inherent structural or grain-alignment characteristics and thus are indicative of non-directional conductivities, whereas results from the well tests are considered indicative of formation conductivity primarily in the horizontal direction. The slug and pump-test results indicate best-value hydraulic conductivities for screened intervals in Memphis aquifer wells for which corresponding grain-size results are available that average about 17 ft/d and range from less than 1 to 100 ft/d (Table 2.4). Differences between the data sets likely are the result of differences in the measurement or estimation techniques and associated factors which are discussed further below.

Hydraulic-conductivity estimates calculated from the Memphis aquifer grain-size data using a modification to the Hazen method (Kasenow, 1997) are based on the d_{10} grain size and thus do not consider the finest fraction of the samples that could tend to reduce the conductivity, particularly in poorly sorted silty sands. Also, grain-size data normally reflect samples of limited size relative to regions of a formation stressed during well tests, and formula-derived estimates using these data may not take into account grain or pore shape, packing, anisotropy, and other variables that can influence hydraulic conductivity. For these reasons, hydraulic conductivity of the Memphis aquifer may tend to be overestimated based on the grain-size analyses of samples from this unit.

Conversely, horizontal hydraulic conductivity values determined from slug and pumping tests of wells screened in unconsolidated sediments can be affected by factors such as formation disturbance during drilling, formation drilling mud (if used) or air invasion, sand/gravel pack emplacement, and adequacy of well development, each of which tend to negatively affect (reduce) actual formation permeability. Rotasonic drilling methods were selected for use during the well installation phase of the project on the basis of the types and depths of the wells and on the ability of this method to collect continuous core from the borings. Differences in hydraulic conductivity estimates between the well tests and grain-size results suggest that the rotasonic drilling technique, coupled with the use of drilling mud in some of the borings to overcome problems with heaving, very fine sand and resulting lack of core recovery in the upper part of the Memphis aquifer, tended to disturb the formation during drilling and reduce the permeability of sediments adjacent to the screened intervals. Furthermore, differences in the hydraulic conductivity between falling and rising-head slug tests conducted in some of the Memphis aquifer wells imply that insufficient development of some of the wells has occurred. All wells installed for the study except one (well 45) were developed by various methods and amounts (Table 2.3); however, caution was taken during this process not to "overdevelop" the wells so that results of water-quality determinations would be negatively influenced. For these reasons, hydraulic conductivity of the Memphis aquifer may tend to be underestimated based on slug and pump tests of wells screened in this unit.

Estimates of hydraulic conductivity based on the grain-size results range from 137 to 324 ft/d (Figures 2.12 and 2.13) for samples of the alluvial (shallow) aquifer at both the confined and unconfined well-cluster locations, with an average of about 230 ft/d for both sites. Slug tests were only conducted on one well screened in the alluvial aquifer (44) during this study. Analysis of the test results from this well produced an estimate of horizontal hydraulic conductivity of 100 ft/d, which compares more favorably with estimates from grain-size samples from this formation than for results from the two different estimation methods used for the Memphis aquifer. Because of the relatively shallow depth of well 44 (50 feet), the dominant sand and gravel composition of the alluvial sediments, and no use of drilling mud during installation of this well, it is likely that the formation was less disturbed by the rotasonic drilling process and has maintained hydraulic characteristics that are more representative of the undisturbed alluvium.

WATER-LEVEL AND WOLF RIVER DATA

As described in Chapter 2, continuous water-level data were collected from selected wells in the Shelby Farms study area during part of the investigation. Comparison of these data and stage data for the Wolf River at the Walnut Grove Road bridge that were estimated from stage data collected at the presently active Wolf River stream gage located at Germantown Road. (Figures 2.16-2.18 demonstrate the influence of the window feature on groundwater/surface-water interaction in the study area.)

Water levels measured in well MS-17 screened in the Memphis aquifer are 25 to 30 feet below those measured in well 19 screened in the alluvial aquifer and the estimated stage of the Wolf River at Walnut Grove Road (Figure 2.16), indicating that the presence of the upper Claiborne confining unit produces hydraulic separation and a relatively large vertical gradient downward between the shallow aquifer/surface-water system and the Memphis aquifer at this location. Water levels measured in well MS-14 screened in the upper part of the Memphis aquifer at the unconfined location are only a few feet lower than those measured in well 44 screened in the alluvial aquifer at this site, indicating the absence of the confining unit produces hydraulic connection and a relatively smaller vertical gradient downward from the shallow aquifer to the Memphis aquifer through the window. Using a head difference of about 6 feet measured between wells 44 and MS-14 at the window location during the period of continuous water-level data collection (Figure 2.18), the difference in well screen depths (about 100 feet) between these wells (Table 2.3), and an estimated range in vertical hydraulic conductivity of about 0.2 to 1.0 ft/d based on an assumed 1:10 ratio between the average vertical and horizontal hydraulic conductivity values for the Memphis aquifer determined from the slug/pump tests (17 ft/d and grain-size analyses (116 ft/d), a groundwater flux of between about 0.01 and 0.06 ft/d is estimated for a 1-ft^2 area of the aquifer at the window location. Calculated average groundwater velocities vertically downward through the window range from about 0.3 to 1.5 ft/d. Thus, estimated lengths of time required for groundwater to move downward from the alluvial aquifer to a depth of about 100 feet below the top of the Memphis aquifer at the window location range from about 330 to 70 days based on the hydraulic data.

Water levels in wells 19 and MS-17, which are screened in the shallow and Memphis aquifers, respectively, and are located about 1,000 feet from the Wolf River (Figure 2.7), vary systematically with Wolf River stage, with the most pronounced effects seen in well 19 (Figure 2.16).

Water levels in the other wells screened in the shallow aquifer (18 and 44) and well MS-14 screened in the Memphis aquifer show little if any correspondence to changes in stage of the Wolf River (Figures 2.17 and 2.18), likely resulting from their greater distances from the river relative to the southwestern well cluster. Based on the distance from the river to the southwestern well cluster and the timing of the responses seen in well 19 in relation to the increased-stage events, corresponding rises in water levels seen in this well may result from a combination of loading effects from the river and recharge to the shallow aquifer from lost stream flow. Because of the confined conditions of the Memphis aquifer at this location, the more subtle water-level rises seen in well MS-17 most likely correlate to and are the result of loading effects on the aquifer from increases in stage of the Wolf River.

Water levels measured in wells screened in the shallow aquifer were lower than the stage of the Wolf River during the period of record (except possibly in well 18 for the period from early July to mid-August 2003; Figure 2.17). Water levels ranged from about 215-217 feet above mean sea level in well 44 near the center of the window, about 13-15 feet lower than the approximately 230-foot altitude of the base-flow stage of the Wolf River at Walnut Grove Road (Figure 2.18), to about 228-231 feet above mean sea level in well 19, which is about 2 feet lower to 1 foot higher than the base-flow altitude of the Wolf River at Walnut Grove Road (Figure 2.18). These data support results of previous investigations (Bradley, 1991; Parks and Mirecki, 1992), which indicate that the water-level gradients and flow directions in the shallow aquifer north of the Wolf River under most conditions are northeast away from the river toward and into the window feature and then downward to the Memphis aquifer.

Bradley (1991) reported that stream-flow discharge measurements made by the USGS in October 1962 under approximate base-flow conditions along the Wolf River at Germantown Road upstream from the present study area and at a site about 2 miles downstream from the study area showed a loss of flow of 13 ft^3/s between the two sites. Additional discharge measurements made by the USGS in November 1986 under base-flow conditions showed a stream-flow loss of about 16 ft^3/s over a 2.1-mile reach of the river from an upstream measurement location and Walnut Grove Road (Bradley, 1991). An earlier set of measurements made by the USGS in August 1986 for the same study over a longer reach of the Wolf River under base-flow conditions indicated a general increase in flow from upstream to downstream sites (Bradley, 1991). As a part of the present study, stream-flow measurements were made by the USGS in July 2002 at three locations on the Wolf River under base-flow conditions (Table 2.5). The measurements showed a small possible loss of stream flow (2 ft^3/s) between Germantown Road and Walnut Grove Road and a slightly larger possible increase in stream flow (5 ft^3/s) from Walnut Grove Road to a measurement site located about 3.5 miles downstream. These data are presumed inconclusive, however, because the differences in the discharge are within ±5 percent of the total flow measured at each site, the maximum error range used by the USGS for determining measurement acceptability. Water-level and stream-flow data collected historically and during this investigation support the general conclusion that the Wolf River loses flow to the alluvial aquifer in or near the study area and that this water supplies a component of the groundwater flow that moves toward and into the window, then down to the Memphis aquifer.

Periodic water-level data collected during this study show the configuration of the water-table in the shallow aquifer and the potentiometric surface of the Memphis aquifer in and near the study area (Figures 2.14 and 2.15). The maps clearly show the depression in the water-table in the alluvial aquifer and corresponding mounding in the potentiometric surface of the Memphis aquifer, which results from leakage through the window. With increased control provided by test holes and wells installed for this study, and by the potentiometric data collected during the study, the area of the window(s) has been further defined, providing for increased confidence of the region through which leakage is occurring. These data have been used to develop the present conceptual hydrostratigraphic and numerical groundwater flow models that are discussed further in the following sections.

HYDROSTRATIGRAPHIC CONCEPTUAL MODEL FOR THE SHELBY FARMS WINDOW

The results from the AwwaRF tailored collaboration project at Shelby Farms has greatly increased understanding of the origin and hydrostratigraphic properties of a window through the upper Claiborne confining unit in the Memphis area. Previous studies of the potential for contamination by leakage through the confining unit (Graham and Parks, 1986; Parks, 1990; Parks and Mirecki, 1992; Bradley, 1991; Kingsbury and Parks, 1993; Parks et al., 1995; Larsen et al., 2003) have typically specified that leakage occurs where the confining unit is thin or absent, but most have not specified the origin of the thinning or absence. Three potential origins for hydrostratigraphic windows in the confining unit include: (1) presence of transmissive sands within otherwise clay-rich confining unit, (2) Post-Claiborne erosion of clay-rich facies and subsequent deposition or valley-filling sand and gravel, and (3) faulting-related disruption and displacement of clay-rich confining unit resulting in a transmissive zone. Although zones of transmissive sands within the upper Claiborne confining unit have been identified in past studies (Parks, 1990; Larsen et al., 2003), the results of the Shelby Farms study are the first to establish clearly that the Memphis aquifer locally extends upward through the upper Claiborne confining unit and is directly overlain by the alluvial deposits. No evidence of a Post-Claiborne paleovalley is observed at the Shelby Farms site; however, examples of paleovalley or erosional windows are noted from the Davis well field (Parks et al., 1995) and Sheahan well field (Larsen et al., 2003). At present, none of the known windows through the confining unit in Shelby is clearly and specifically associated with a fault.

The results of the Shelby Farms study also showed that the sediments comprising the upper Claiborne confining unit vary widely in their properties and display complex lateral relationships. The two facies (Facies A and B, Figure 2.27) identified in the confining unit vary in color, grain-size, density, mineralogy and organic content, suggesting differing hydraulic behaviors. It is not clear from the present study how common these two facies are within the Memphis aquifer, but both have been identified in field exposures within the Memphis area. Research is currently in progress to further clarify the extent and stratigraphic relationships of Facies A and B within the greater Memphis area.

The presence of windows associated with sedimentary heterogeneities in the confining unit, such as that observed at Shelby Farms, is interpreted to be related to processes during deposition of the Claiborne Group (which includes the Memphis aquifer and confining unit). Current data indicate that the Claiborne Group sediments deposits in the Memphis area were deposited in a series of ancient deltas, coastal swamps, and low-gradient streams that lined the ancient Mississippi Embayment basin. During sea-level highstands, fine-grained deltaic sediment with intervening channel sands may have been deposited broadly around the Embayment. Confining unit facies A is interpreted to reflect this style of sedimentation. However, during subsequent sea-level lowstands, sedimentation would have shifted largely to downdip ancestral Gulf Coast locations or in the center of the Embayment. During sea-level lowstands substantial incision may occur into the previously deposited sediments, creating paleovalleys in updip regions. Such ancient paleovalleys can extend hundreds of miles upgradient of ancient coastlines (e.g., Plint and Wadsworth, 2003). During rise of sea level these paleovalleys are often filled with fluvial and estuarine sands and clays prior to more widespread deposition of sediment. Confining unit facies B and associated sandy channel deposits are interpreted to represent this style of sedimentation. The Memphis aquifer sediments are interpreted to reflect fluvial deposition further updip of the coastal plain region, but the sandy window deposits attributed to the Memphis aquifer at Shelby Farms (Figure 2.27) may be a sandy channel deposit associated with confining unit B.

The degree to which facies A versus facies B is preserved within an area is largely dependent upon tectonic subsidence within the Mississippi Embayment and sediment supply. Assuming a steady sediment supply and waning subsidence in the Mississippi Embayment during Middle Eocene time (consistent with basin model of Cox and Van Arsdale, 1997), updip incision of high-stand deposits would have likely occurred and led to facies B deposition within the confining unit, especially eastward of the Eocene depocenter of the Mississippi Embayment along the trace of the modern Mississippi River. Similar types of erosional features in updip fine-grained marine and nonmarine deposits are observed in Eocene coastal plain strata in South Carolina (Castle and Miller, 2000). The Shelby Farms study and that of Castle and Miller provide case examples of window features that may be more common than are currently recognized.

INTERPRETATION OF ENVIRONMENTAL TRACER RESULTS

The $^3H/^3He$ and geochemical data provide useful information for evaluating the rate of groundwater recharge through the window in the confining unit at the Shelby Farms site. The simplest interpretation of a conservative groundwater tracer flowing through porous media is that of piston flow (Cook and Böhlke, 2000). Given that evidence for mixing of older and younger waters occurs only in the deeper wells (Table 3.3), the piston flow model is a useful approximation for interpreting the downward recharge through the window. In such a case, the tracer moves at the same rate as the groundwater and does not mix appreciably along hypothetical flow tubes. Interpretation of two residence times along a flow tube, assuming piston flow, yields an instantaneous Darcy velocity or if the effective porosity is applied to vertical downward flow, a recharge rate. In the case of $^3H/^3He$ groundwater residence times of water moving vertically through a window in the confining unit, a model recharge rate (R) can be determined using:

$$R = (z_t \times n_e) / (t_2 - t_1) \quad (5.1)$$

where z_t is the distance between points of known residence time, n_e is effective porosity, and t_1 and t_2 are $^3H/^3He$ residence times. Similarly, a Darcy velocity for groundwater moving through a confined aquifer along a flow line can be determined using:

$$v = z_t / (t_2 - t_1) \quad (5.2)$$

Recharge rates and Darcy velocities are calculated from $^3H/^3He$ data using equations (5.1) and (5.2), and are given in Table 5.1. A 10% error was propagated through the calculations in order to reflect uncertainty in the flow rates due to error in age estimates. The Darcy velocities determined from the unconfined site to down gradient wells are viewed as minimum values because the tracer ages reflect only the young water portion of the groundwater mixture. The

Table 5.1
Recharge rates and Darcy velocities calculated from $^3H/^3He$ and hydraulic data

Sampling event	Wells	Distance (ft)	Tracer rate or velocity (ft/day)*	Hydraulic rate (ft/day)†	Type of calculation
November 2002					
	44 and MS-12	30	0.0046 – 0.0037	1.12 – 0.005	Recharge
	44 and MS-13	65	0.0089 – 0.0073	0.67 – 0.060	Recharge
	44 and MS-14	99	0.0061 – 0.0050	1.59 – 0.014	Recharge
	MS-14 and MS-13	34	0.0046 – 0.0027	3.31 – 0.005	Recharge
	MS-12 and MS-13	35	-0.046 – +0.015	0.29 – 0.002	Recharge
	MS-12 and MS-14	69	0.0084 – 0.0052	1.78 – 0.015	Recharge
	MS-14 and MS-17	2282	3.59 – 0.57	0.43 – 0.191	Darcy velocity
	MS-14 and MS-18	3655	1.48 – 0.59	0.45 – 0.006	Darcy velocity
	MS-14 and MS-16	1243	0.44 – 0.19	0.33 – 0.010	Darcy velocity
June 2003					
	44 and MS-13	65	0.0072 – 0.0055	0.82 – 0.004	Recharge
	44 and MS-14	99	0.013 – 0.010	1.76 – 0.157	Recharge
	MS-14 and MS-13	34	-0.01 – +0.034	3.56 – 0.032	Recharge
	MS-14 and MS-17	2282	1.04 – 0.50	0.44 – 0.196	Darcy velocity
	MS-14 and MS-18	3655	1.25 – 0.67	0.47 – 0.006	Darcy velocity
	MS-14 and MS-16	1243	0.34 – 0.20	0.38 – 0.012	Darcy velocity

* n_e assumed to be 0.40; age date assumed to have 10% error.

† n_e assumed to be 0.40; range of values reflects lab versus field-determined hydraulic conductivity values; lab values are geometric means calculated from grain-size data, field values are best determined values from well tests (Table 2.4); vertical hydraulic conductivity assumed to be one-tenth of lab or field values; equivalent vertical and horizontal hydraulic conductivities calculated using method of Leonards (1962).

Darcy velocities calculated between the unconfined and confined sites best approximate flow along a flow line; the velocities between the unconfined sites and MS-17 and MS-18 are given to approximate deviations in the flow direction. Recharge rates and Darcy velocities calculated using hydraulic head data obtained during the sampling events are also given in Table 5.1. The hydraulic conductivity determined from both grain-size data and well tests were used to bracket the range of velocities. The geometric mean of the grain-sized based conductivities for each hydrostratigraphic unit was used to calculate the overall hydraulic conductivity between well screens or well locations. Only grain-size data from the unconfined and confined well clusters were used; values in the Memphis aquifer in wells MS-17 and -18 were approximated using data from the confined cluster. The grain-size based rates are in all cases faster than the well test-based rates, probably reflecting semi-consolidation of the materials which is not accounted for in the grain-size based estimates.

The recharge rates calculated using the tracer data are two to three orders of magnitude less than those determined using spatially averaged recharge rate of 0.45 ft/d from groundwater flow modeling Experiment One (Table 4.2) or values determined from grain-size based Darcy's law calculations (Table 5.1). The data are more consistent with the spatially averaged recharge rate of 0.023 ft/d obtained from groundwater flow modeling experiment Two (Table 4.4) and nearly overlap with the well-test based Darcy's law calculations (Table 5.1). The tracer-based recharge rates between MS-12 and MS-13 for the November 2002 sampling event and MS-13 and -2 for the June 2003 sampling event indicate potential for downward or upward flow. It is not clear whether this is a function of overlapping age distributions between closely spaced well screens or reflects actual reversal in flow direction.

The tracer-based Darcy velocities between the unconfined site (MS-14) and the confined site (MS-16) overlap with the grain-size based Darcy velocities, but are an order of magnitude greater than the well-test based Darcy velocities. Because the waters in the deeper wells, which include MS-14 and -16, are mixtures of young recharge water and older Memphis aquifer water, the tracer-based velocities are expected to be greater than the actual values. Thus, these results are consistent with the recharge estimates and suggest that the well-test based Darcy velocity calculations are reasonable.

The residence times and recharge rates calculated for the unconfined cluster from the November 2002 data indicate increasing age with depth, consistent with typical piston-flow models of recharge to an aquifer (Cook and Böhlke, 2000). The geochemical data and mixing relationships (Table 3.6) indicate that the well with the greatest leachate proportion is MS-14, which is contrary to what would be expected with dispersion acting along the flow path. However, if one assumes that leachate contributions declined greatly after the Shelby County landfill was capped in 1988, then the higher leachate contributions in MS-14 might be expected to reflect migration of leachate over approximately 15 years since then and, thus, would be more consistent with the residence time calculations (19.8 yrs for MS-14). Although water quality in an alluvial well down gradient from the landfill (Sh:Q-129) did improve between sampling events in 1990 (data from Parks and Mirecki, 1992) and 2000, water quality in well MS-12 remained approximately the same between the 1990 and 2002 sampling events.

Comparison of residence times and recharge rates from the November 2002 and June 2003 sampling events suggests that the flow conditions are more complex than that of a simple piston flow model. For a piston flow model with ages ranging up to 20 years, the residence times should not vary appreciably between seasons; however, the residence times determined for the June 2003 event are as much as 10 years younger than those for the November 2002 event. The fact that younger residence times are determined for the June event suggests increased recharge of younger water during this time, as was suggested based on the reconstructed ^3H loadings (Table 3.3). However, little evidence for an appreciable increase in the water table is observed between the fall and winter months (Figure 2.14). Geochemical data were not obtained from the June 2003 sampling event so it is currently unclear whether water-quality changes accompanied the younger residence times obtained in Spring 2003.

One possible explanation for the combined water quality and residence time variations is recharge from either an elongate window (elongated toward the landfill) or multiple small windows in an array extending from the landfill to the unconfined well cluster. Evidence supporting this interpretation includes observation of the lowest water-table elevations centered

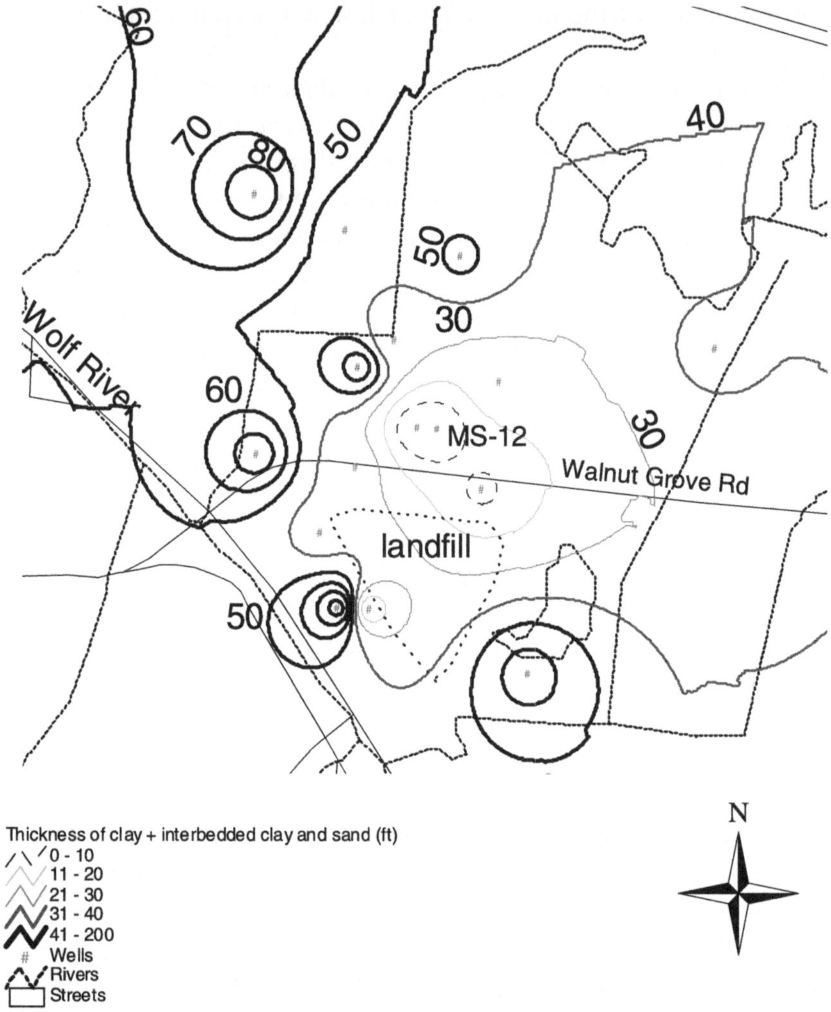

Figure 5.1 Isopach map of total clay thickness in area of Shelby Farms window site

on well Sh:Q-119 (along the northern border of the landfill) (Figure 2.14), the isopach map of clay thickness (Figure 5.1) showing minima extending toward the landfill from MS-12, water quality impacts from leachate in Memphis aquifer wells near the landfill (Parks and Mirecki, 1992; Mirecki and Parks, 1994) and groundwater mounding near the landfill from the groundwater flow model results (Figure 4.5).

Variations in residence times measured at the unconfined well cluster could reflect subtle changes in the amount of mixing between upgradient Memphis aquifer water, upgradient leakage points near the landfill, and leakage at the unconfined cluster. The high leachate proportion present in well MS-14 could reflect leakage near the landfill and subsequent flow downward and lateral toward the unconfined cluster as illustrated in Figure 5.2. Further tracer and geochemical data are needed to quantitatively test this model but the present results qualitatively fit the model.

The geochemical results (Table 3.5) from wells down gradient from the unconfined well cluster provide information regarding the local extent of landfill leachate impacts upon water quality in the Memphis aquifer and flow paths downgradient from the window location. Landfill leachate impacts groundwater quality by increasing total dissolved solids, trace metals, and

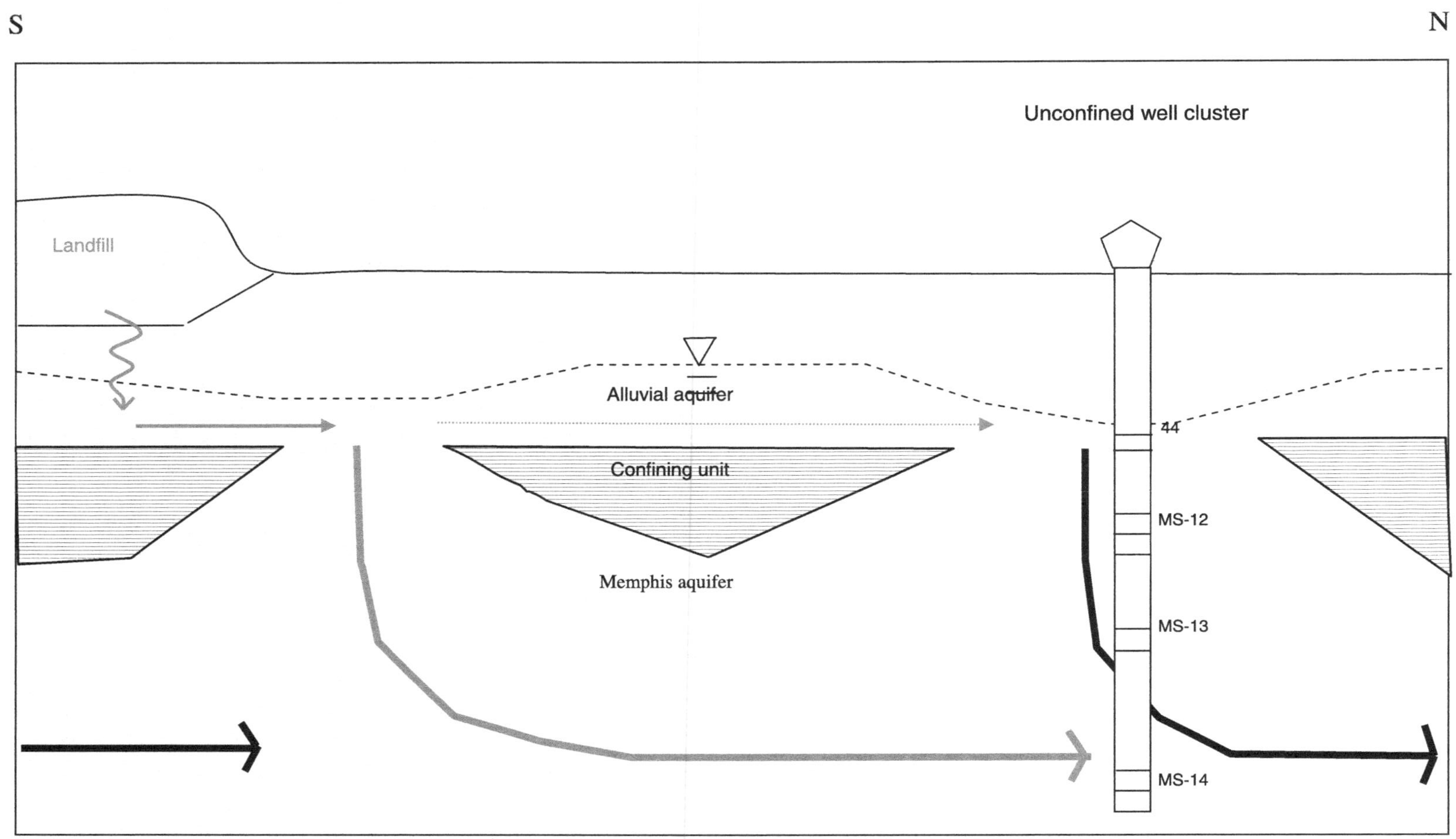

Figure 5.2 Diagram illustrating a possible configuration of windows in the confining unit and resulting effects on water quality at the Shelby Farms site

dissolved organic carbon contents, and by creating local zones of highly reduced (typically methanogenic) redox conditions (Baedecker and Back, 1979a, b; Lyngkilde and Christensen, 1992; Kehew, 2001). The groundwater quality impacts of leachate vary tremendously depending on local hydrogeology, type of waste disposed, capping and liner technology implanted, and age. Parks and Mirecki (1992) documented increased dissolved major and trace inorganic concentrations over background levels within samples from wells in the Memphis aquifer downgradient from the landfill. Total organic carbon contents were variable in the affected Memphis aquifer wells, but low concentrations of 18 synthetic organic carbon compounds were detected in eight wells screened in the Memphis aquifer. Mirecki and Parks (1994) used barium and strontium concentrations to estimate the leachate impacts within the Memphis aquifer and determined that as much as 37% (based on strontium mixing relationship) of the dissolved mass in water from well MS-12 was derived from leachate.

The results of the present study demonstrate landfill leachate impacts within the Memphis aquifer extending to all Memphis aquifer wells (MS-15, -16, -17, and -18) downgradient from the unconfined cluster. Furthermore, the maximum impacts of landfill leachate were previously identified in well MS-12 (Mirecki and Parks, 1994); however, the results of the present study indicate greater impact at depth at the unconfined site (wells MS-13 and -14) and impacts similar to that of MS-12 at the downgradient wells (MS-15, -16, -17, and -18). Although methane was not analyzed in the present study, the low concentration and absence of sulfate in wells MS-13 and -14, respectively, indicate an extensive reduced leachate plume in the Memphis aquifer. The low redox potential measured in well MS-16 and presence of sulfide in well MS-15 suggest that the low redox effects extend toward the wells downgradient of the unconfined site. The elevated sulfate and nitrate concentrations in well MS-18 may reflect oxidation of sulfur and nitrogen species, respectively, at the downgradient edge of the leachate impacted water.

INTERPRETATION OF MODELING RESULTS

The modeling analyses by Gentry, et al. (2003) and in Experiment One of the numerical simulation herein considered the recharge flux and total flow balance in the aquifer to be a function of the full transmissivity of the Memphis aquifer. In order to affect the mounding demonstrated by the observed heads at the study site, substantial leakage rates would be necessary (Table 4.2). Further, considering the full transmissivity of the aquifer, the genetic algorithm analyses (Figure 4.6) demonstrate that lower recharge rates result in a probable leakage area that is not consistent with the findings of the hydrostratigraphic conceptual model (Figure 2.27). The recharge rates required to effect the necessary heads with a hypothesis of reduced transmissivity was demonstrated by Experiment Two of the numerical simulation (Table 4.5 and Figure 4.8), and found to be more consistent with recharge values calculated by environmental tracers (Table 5.1). Considering model error, a global reduction results in a better simulation of the observed heads in the Memphis aquifer. These findings suggest that the upper Memphis aquifer may be hydraulically independent of the deeper zones within the localized study area.

The modeling results do not provide a definitive areal extent of the window at the study site. The preliminary genetic algorithm modeling used a constant recharge rate with a possibility of being active or in-active at any node in the search area (Figure 4.3), whereas the MODFLOW 2000 numerical experiments allowed the selection of variable recharge rates for six search zones (Figure 4.4). Both techniques favored higher recharge rates toward the general direction of the landfill (Figures 4.3, 4.6 and 4.8). This may indicate a geometric orientation or elongation of the window toward the landfill as suggested previously.

The conclusions of the modeling analyses support a recharge rate of 0.09 ft/day, or lower, (Table 4.5) with a likely volumetric flux on the order of 36×10^3 ft^3/day. It is concluded that the upper Memphis aquifer behaves independently of deeper zones, resulting in a lower effective transmissivity, which may be characteristic of the Memphis aquifer.

CHAPTER 6
SUMMARY OF OUTREACH ACTIVITIES

A number of conference abstracts and publication efforts have resulted either directly or indirectly from the research performed herein. A summary of those efforts is provided as follows.

JOURNAL PUBLICATIONS

Gentry, R.W., Larsen, D., Ivey, S.S. 2003. Efficacy of Genetic Algorithm to Investigate Small Scale Aquitard Leakage, ASCE Journal of Hydraulic Engineering, V129, n7, pp. 527-535.

CONFERENCE ABSTRACTS

Garner, C.B., Larsen, D., Carmichael, J., and Gentry. R. 2003, Hydrostratigraphy of a Window Through the Upper Claiborne Confining Unit, Memphis, Tennessee. Geological Society of America Abstracts with Programs, vol. 35, no. 1, p. 27.

Gentry, R.W. 2002. The Use of Genetic Algorithms as an Inverse Technique to Guide the Design and Implementation of Research at a Test Site in Shelby County, Tennessee, Eos Trans. AGU, 83(47), Fall Meet. Suppl., Abstract H51C-05.

Gentry, R.W., McKay, L.D., Larsen, D., Carmichael, J.K., Solomon, D.K., Thonnard, N., and Anderson, J.L. 2003. Inter-aquifer Dynamics in and near a Confining Unit Window in Shelby County, Tennessee, USA, Eos Trans. AGU, 84(47), Fall Meet. Suppl., Abstract H21D-0868.

Kenst, A. 2003. Movement of Viruses in the Vadose Zone: A Review of Transport Mechanisms. TNAWRA Symposium, April 9-11, 2003, Burns, TN.

INVITED PRESENTATIONS

Gentry, R.W. The Use of Environmental Tracers to Better Understand Inter-Aquifer Hydraulics in Shelby County, Tennessee, AwwaRF Technology Transfer Conference, Miami, January 25, 2001.

The researchers will continue to disseminate the results of this research in peer-reviewed journals and have held preliminary discussions planning those efforts.

REFERENCES

Aeschbach-Hertig, W., F. Peeters, U. Beyerle, and R. Kipfer. 2000. Palaeotemperature Reconstruction From Nobel Gases in Ground Water Taking Into Account Equilibration With Entrapped Air. *Nature*, 405:1040–1044.

Baedecker, M.J., and W. Back. 1979a. Hydrogeologic Processes and Chemical Reactions at a Landfill. *Ground Water*, 17:429–437.

Baedecker, M.J., and W. Back. 1979b. Modern Marine Sediments as a Natural Analog to the Chemically Stressed Environment of a Landfill. *Journal of Hydrology*, 43:393–414.

Barrow, J.C. 1994. The Resonant Sonic Drilling Method: An Innovative Technology for Environmental Restoration Programs. *Ground Water Monitoring and Remediation*, 2:153–160.

Bayer R., P. Schlosser, G. Bonisch, H. Rupp, F. Zaucker, and G. Zimmek. 1989. Performance and Blank Components of a Mass Spectrometric System for Routine Measurements of Helium Isotopes and Tritium by the ^{3}He In-growth Method. *Sitzungsberichte der Heidelberger Akademie der Wissenschaften, Mathematisch-naturwissenchaftliche Klasse*, 5:241-279.

Boggs, S., Jr. 2001. *Principles of Sedimentology and Stratigraphy.* Prentice-Hall, Upper Saddle River, NJ.

Bouwer, H., and R.C. Rice. 1976. A Slug Test for Determining Hydraulic Conductivity of Unconfined Aquifers With Completely or Partially Penetrating Wells. *Water Resources Research*, 12(3):423–428.

Bradley, M.W. 1988. Construction, *Geologic, and Ground-Water Data for Observation Wells Near the Shelby County Landfill, Memphis, Tennessee.* Open File Report 88–486. U.S. Geological Survey.

Bradley, M.W. 1991. *Ground-Water Hydrology and The Effects of Vertical Leakage and Leachate Migration on Ground-Water Quality Near the Shelby County Landfill, Memphis, Tennessee.* Water-Resources Investigations Report 90–4075. U.S. Geological Survey.

Brahana, J.V., and R.E. Broshears. 2001. *Hydrogeology and Ground-Water Flow in the Memphis and Fort Pillow Aquifers in the Memphis Area, Tennessee.* Water-Resources Investigations Report 89-4131. U.S. Geological Survey.

Brahana, J.V., W.S. Parks, and M.W. Gaydos. 1987. *Quality of Water from Freshwater Aquifers and Principal Well Fields in the Memphis Area, Tennessee.* Water-Resources Investigations Report 87-4052. U.S. Geological Survey.

Brown, M.C. 1993. A Study of the Aquifer System at the Davis Well Field. Master's thesis. Memphis State University, Tennessee.

Carmichael, J.K., W.S. Parks, J.A. Kingsbury, and D.E. Ladd. 1997. Hydrogeology and Ground-Water Quality at the Naval Support Activity Memphis, Near Millington, Tennessee. Water-Resources Investigations Report 97-4158. U.S. Geological Survey.

Carroll, D. 1997. D.L. Carroll's FORTRAN Genetic Algorithm Driver [Online]. Available: <http://www.uiuc.edu/~carroll/ga.html> [cited January 6, 1997]

Castle, J.W., and R.B. Miller. 2000. Recognition and Hydrologic Significance of Passive-Margin Updip Sequences: An Example from Eocene Coastal-Plain Deposits, U.S.A. *Journal of Sedimentary Research*, 70:1290–1301.

Castro, M.C., A. Jambon, and G. de Marsily. 1998a. Nobel Gases as Natural Tracers of Water Circulation in the Paris Basin. 1. Measurements and Discussion of Their Origin and Mechanisms of Vertical Transport in the Basin. *Water Resources Research*, 34(10):2443–2466.

Castro, M.C., P. Goblet, and E. Ledoux. 1998b. Noble Gases as Natural Tracers of Water Circulation in the Paris Basin. 2. Calibration of a Groundwater Flow Model Using Noble Gas Isotope Data. *Water Resources Research*, 34(10):2467–2483.

Clarke, W.B., W.J. Jenkins, and Z. Top. 1976. Determination of Tritium by Mass Spectrometric Measurements of ^{3}He. *International Journal of Applied Radioactive Isotopes*, 27:515–522.

Clifton, H. E., R.E. Hunter, and R.L. Phillips. 1971. Depositional Structures and Processes in the Non-Barred High-Energy Nearshore. *Journal of Sedimentary Petrology*, 41:651–670.

Coleman, J.M. 1982. *Deltas: Processes and Models of Deposition for Exploration.* Second Edition. Boston: Louisiana State University, Coastal Studies Institute.

Coleman, J.M., and D.B. Prior. 1980. Deltaic Sand Bodies. AAPG Short Course, Education Course Notes Series No. 15.

Cook, P.G., and J.K. Böhlke. 2000. Determining Timescales for Groundwater Flow and Solute Transport. In *Environmental Tracers in Subsurface Hydrology.* Edited by P.G. Cook and A.L. Herczeg. Boston: Kluwer.

Cooper, H.H., and C.E. Jacob. 1946. A Generalized Graphical Method for Evaluating Formation Constants and Summarizing Well Field History. *American Geophysical Union Transactions*, 27:526–534.

Cox, R.T., and R.B. Van Arsdale. 1997. Hotspot Origin of the Mississippi Embayment and Its Possible Impact on Contemporary Seismicity. *Engineering Geology*, 46:201–216.

Craig, H., K. Marti, and R. Wiens. 1993. *A Static Mass Spectrometer With Triple Collection for Nitrogen and Neon Isotopes.* SIO Reference Series #93-11. Scripps Oceanographic Institute, La Jolla, CA.

Cross, B. 1989. *Tritium Deposition Data-Access Program, DEP.BAS.* Reston, Va.: U.S. Geological Survey.

Frederiksen, N.O., L.M. Bybell, R.A. Christopher, A.J. Crone, L.E. Edwards, T.G. Gibson, J.E. Hazel, J.E. Repetski, D.P. Russ, C.C. Smith, and L.W. Ward. 1982. Biostratigraphy and Paleoecology of Lower Paleozoic, Upper Cretaceous, and Lower Tertiary Rocks in U.S. Geological Survey New Madrid Test Wells, Southeastern Missouri Tulane. *Studies in Geology and Paleontology*, 17(2):23–45.

Friedman, G.M., and K.G. Johnson. 1982. *Exercises in Sedimentology.* John Wiley and Sons, New York.

Garner, C.B., D. Larsen, J. Carmichael, and R. Gentry. 2003. Hydrostratigraphy of a Window Through the Upper Claiborne Confining Unit, Memphis, Tennessee. *Geological Society of America Abstracts With Programs*, 35(1):20.

Gee, G.W., and J.W. Bauder. 1986. Particle-Size Analysis. In *Methods of Soil Analysis, Part 1: Physical and Mineralogical Methods.* Edited by Klute. Madison, Wis.: Soil Science Society of America.

Gentry, R.W., C.V. Camp, and J.L. Anderson. 2001. Use of GA to Determine Areas of Accretion to Semiconfined Aquifer. *ASCE Journal of Hydraulic Engineering*, 127(9):738–746.

Gentry, R.W., and D. Larsen. 2001. Final Report for TNWRRC Project—An Investigation to Identify Sources and Quantities of Modern Recharge to the Memphis Aquifer in the Sheahan Well Field in Shelby County, Tennessee, submitted to TN Water Resources Research Center, University of Tennessee.

Gentry, R.W., D. Larsen, and S.S. Ivey. 2003. Efficacy of Genetic Algorithm to Investigate Small Scale Aquitard Leakage. *ASCE Journal of Hydraulic Engineering*, 129(7):527–535.

Graham, D.D., and W.S. Parks. 1986. Potential for Leakage Among Principal Aquifers in the Memphis Area, Tennessee. Water-Resources Investigations Report 85-4295. U.S. Geological Survey.

Greenwood, B., and P.R. Mittler. 1987. Vertical Sequence and Laterial Transitions in the Facies of a Barred Nearshore Environment. *Journal of Sedimentary Petrology*, 55:366–375.

Halford, K.J., and E.L. Kuniansky. 2002. Documentation of Spreadsheets for the Analysis of Aquifer-Test and Slug Test Data. Open-File Report 2002-197. U.S. Geological Survey.

Harbaugh, A.W., E.R. Banta, M.C. Hill, and M.G. McDonald. 2000. MODFLOW-2000, the U.S. Geological Survey Modular Ground-Water ModelÑUser Guide to Modularization Concepts and the Ground-Water Flow Process. Open-File Report 2000-92. U.S. Geological Survey.

Hays, P.D., J.K. Lovelace, and T.B. Reed. 1998. Simulated Response to Pumping Stress in the Sparta Aquifer of Southeastern Arkansas and North-Central Louisiana, 1998–2027. Water-Resources Investigations Report 98-4121. U.S. Geological Survey.

Ivey, S. 2003. The Development of a Combined Age Distribution and Inverse Technique to Identify Wellhead Protection Zones for A Semi-Confined Aquifer. Ph.D. dissertation. University of Memphis, Tennessee.

Kasenow, M. 1997. *Applied Ground-Water Hydrology and Well Hydraulics.* Highlands Ranch, Colo.: Water Resources Publications.

Kehew, A.E. 2001. *Applied Chemical Hydrogeology.* Upper Saddle River, NJ: Prentice-Hall.

Kenley, D. 1993. A Ground Water Flow Model of the Northern Mississippi Embayment. Master's thesis. Memphis State University, Tennessee.

Kingsbury, J.A., and W.S. Parks. 1993. Hydrogeology of the Principal Aquifers and Relation of Faults to Interaquifer Leakage in the Memphis Area, Tennessee. Water-Resources Investigations Report 93-4075. U.S. Geological Survey.

Larsen, D., R.W. Gentry, S. Ivey, D.K. Solomon, and J. Harris. 2003. Groundwater Leakage Through a Confining Unit Beneath a Municipal Well Field, Memphis. Tennessee, USA. In *Geochemical Processes in Soil and Groundwater.* Edited by H.D. Schulz and A. Hadeler. Berlin: Wiley-VCH.

Leonards, G.A. 1962. *Engineering Properties of Soils, Foundation Engineering*. Edited by G.A. Leonards. New York: McGraw-Hill.

Lyngkilde, J., and T.H. Christensen. 1992. Redox Zones of a Landfill Leachate Plume. *Journal of Contaminant Hydrology*, 10:273–289.

Mirecki, J.E., and W.S. Parks. 1994. Leachate Geochemistry at a Municipal Landfill, Memphis, Tennessee. *Ground Water*, 32(3):390–398.

Moore, G.K., and D.L. Brown. 1969. Stratigraphy of the Fort Pillow Test Well Lauderdale County, Tennessee. Report of Investigations 26. Tennessee Division of Geology.

Ng, Y.G. 1993. Ground Water Modeling at the Shelby County Landfill. Master's thesis. Memphis State University, Tennessee.

Outlaw, J.E. 1994. A Ground Water Flow Analysis of the Memphis Sand Aquifer in the Memphis, Tennessee, Area. Master's thesis. University of Memphis, Tennessee.

Parks, W.S. 1990. Hydrogeology and Preliminary Assessment of the Potential for Contamination of the Memphis Aquifer in the Memphis Area, Tennessee. Water-Resources Investigations Report 90-4092. U.S. Geological Survey.

Parks, W.S., and J.K. Carmichael. 1990. Geology and Ground-Water Resources of the Memphis Sand in Western Tennessee. Water-Resources Investigations Report 88-4182. U.S. Geological Survey.

Parks, W.S., and J.E. Mirecki. 1992. Hydrogeology, Ground-Water Quality and Potential for Water-Supply Contamination Near the Shelby County Landfill in Memphis, Tennessee. Water-Resources Investigations Report 91-4173. U.S. Geological Survey.

Parks, W.S., J.E. Mirecki, and J.A. Kingsbury. 1995. Hydrogeology, Ground-Water Quality, and Source of Ground Water Causing Water-Quality Changes in the Davis Well Field at Memphis, Tennessee. Water-Resources Investigations Report 94-4212. U.S. Geological Survey.

Plint, A.G., and J.A. Wadsworth. 2003. Sedimentology and Palaeogeomorphology of Four Large Valley Systems Incising Delta Plains, Western Canada Foreland Basin: Implications for Mid-Cretaceous Sea-Level Changes. *Sedimentology*, 50:1147–1186.

Richardson, G.J. 1989. A Study of Potential Sources of Leakage into the Memphis Sand Aquifer Beneath the Davis Well Field in Memphis, Tennessee. Master's thesis. Memphis State University, Tennessee.

Robinson, J.L., J.K. Carmichael, K.J. Halford, and D.E. Ladd. 1997. Hydrogeologic Framework and Simulation of Ground-Water Flow and Travel Time in the Shallow Aquifer System in the Area of Naval Support Activity Memphis, Millington, Tennessee. Water-Resources Investigations Report 97-4228. U.S. Geological Survey.

Shapiro, A.M., J. Margolin, S. Dolev, and Y. Ben-Israel. 1997. A Graphical-User Interface for the U.S. Geological Survey Modular Three-Dimensional Finite-Difference Ground-Water Flow Model (MODFLOW-96) Using Argus Numerical Environments. Open-File Report 97-121. U.S. Geological Survey.

Smethie, W.M., D.K. Solomon, S.L. Schiff, and G.D. Mathieu. 1992. Tracing Groundwater Flow in the Borden Aquifer Using Krypton-85. *Journal of Hydrology*, 130:279–298.

Solomon, D.K., and P.G. Cook. 2000. ^{3}H and ^{3}He. In *Environmental Tracers in Subsurface Hydrology*. Boston: Kluwer.

Solomon, D.K., R.J. Poreda, S.L. Schiff, and J.A. Cherry. 1992. Tritium and Helium 3 as Groundwater Age Tracers in the Borden Aquifer. *Water Resources Research*, 28(3):741–756.

Stute, M., and P. Schlosser. 2000. Atmospheric Noble Gases. In *Environmental Tracers in Subsurface Hydrology*. Boston: Kluwer.

Thonnard, N., and B.L. Lehmann. 1995. The Institute for Rare Isotope Measurements: An Inter-Disciplinary Facility for RIS-based Applications. AIP Conference Proceedings.

Thonnard, N., and L.D. McKay. 1997. Emerging New Ultra-Sensitive Analytical Techniques Using Resonance Ionization: Applications in Groundwater Dating and Other Trace Analyses. ATV 1997. Committee on Groundwater Pollution, Danish Academy of Technical Sciences, 135–149, April 23.

Thonnard, N., M.C. Wright, W.A. Davis, and R.D. Willis. 1992. *The Second-Generation RIS-TOF Noble Gas Detector: Detection Limits Below 100 Atoms in Less Than 5 Minutes*. Institute of Physics Conference Series.

U.S. Geological Survey. 1993. Ellendale quadrangle, Tennessee [map]. 1:24,000. 7.5 Minute Series. Washington, DC: USGS.

ABBREVIATIONS

AMSL	above mean sea level
AW	archived as water in hermetic glass sampling bottle
AWWA	American Water Works Association
AwwaRF	Awwa Research Foundation
BLS	below land surface
BP	barometric pressure
DD	drawdown
DG	degassed and stored in 4.8-liter hermetic aluminum cylinder
DOE	Department of Energy
EM	electromagnetic
GA	genetic algorithm
gpm	gallons per minute
IRIM	Institute for Rare Isotope Measurements
MLG&W	Memphis Light, Gas & Water
N/A	not applicable
NSF	National Science Foundation
OD	outer diameter
PVC	polyvinyl chloride
REC	recovery
TOC	top of casing
TU	tritium unit
U.S.	United States
USGS	United States Geological Survey

IWA Publishing's authorised EU representative for General Product Safety Regulations is Diane D'Arras, 15 rue Duret, 75116 Paris, France, e-mail: safety@iwap.co.uk.

Printed and bound by CPI Group (UK) Ltd, Croydon, CR0 4YY
17/09/2026
02218798-0010